6th International Conference on Mechatronics and Control Engineering (ICMCE 2017)

Journal of Physics: Conference Series Volume 1016

St. Petersburg, Russia
2 – 4 December 2017

ISBN: 978-1-5108-6321-7
ISSN: 1742-6588

Printed from e-media with permission by:

Curran Associates, Inc.
57 Morehouse Lane
Red Hook, NY 12571

Some format issues inherent in the e-media version may also appear in this print version.

Printed by Curran Associates, Inc. (2018)

For permission requests, please contact the Institute of Physics
at the address below.

Institute of Physics
Dirac House, Temple Back
Bristol BS1 6BE UK

Phone: 44 1 17 929 7481
Fax: 44 1 17 920 0979

techtracking@iop.org

Additional copies of this publication are available from:

Curran Associates, Inc.
57 Morehouse Lane
Red Hook, NY 12571 USA
Phone: 845-758-0400
Fax: 845-758-2633
Email: curran@proceedings.com
Web: www.proceedings.com

Table of contents

Volume 1016

6th International Conference on Mechatronics and Control Engineering (ICMCE 2017)

2–4 December 2017, St. Petersburg, Russian Federation

Accepted papers received: 20 April 2018
Published online: 21 May 2018

Preface

6th International Conference on Mechatronics and Control Engineering (ICMCE 2017)

Conference Committee

Peer review statement

Papers

Development and performance analysis of functional materials

Modeling and Sound Insulation Performance Analysis of Two Honeycomb-hole Coatings

H F Ye, M Tao and W Z Zhang.....1

Composite adhesive bonds reinforced with microparticle filler based on egg shell waste

Miroslav Müller and Petr Valášek.....6

Degradation of strength properties of epoxy resin filled with natural-based particles

Petr Valášek, Karolína Habrová and Miroslav Müller.....11

Experience of light thin-walled structures improvement in construction

A V Frolovskaia, S V Deordiev, A Falk, N Y Klinduh and I I Terehova.....16

Robot design and control system

Control of Sit to Stand Mechanism of Assistive Device for Paraplegics

A Abdullah and Z Kausar.....21

A spatial registration method for navigation system combining O-arm with spinal surgery robot

H Bai, G L Song, Y W Zhao, X Z Liu and Y X Jiang.....26

Modelling and Closed-Loop System Identification of a Quadrotor-Based Aerial Manipulator

Chioniso Dube and Jimoh O Pedro.....33

Trajectory tracking control for a nonholonomic mobile robot under ROS

Khadir Lakhdar Besseghieur, Radosław Trębiński, Wojciech Kaczmarek and Jarosław Panasiuk.....42

Output Tracking of Some Classes of Non-minimum Phase Nonlinear Systems By Redefinition Output Approach

Janson Naiborhu, Firman and Edwin Setiawan.....47

Parametric synthesis of a robust controller on a base of mathematical programming method

I.V. Khozhaev, S.A. Gayvoronskiy and T.A. Ezangina.....53

An improved measurement system for FOG pure lag time with no changing of FOG work status

X Chen, J H Yang, Y L Zhou and X W Shu.....58

Distributed Nash Equilibrium Seeking for Generalized Convex Games with Shared Constraints

Chao Sun and Guoqiang Hu.....66

Mechanical engineering and dynamic analysis

Research on Equivalent Tests of Dynamics of On-orbit Soft Contact Technology Based on On-Orbit Experiment Data

F Yang, Z H Dong and X Ye.....75

Research on The Construction of Flexible Multi-body Dynamics Model based on Virtual Components

Z H Dong, X Ye and F Yang.....81

Simulation study on dynamics model of two kinds of on-orbit soft-contact mechanism

X Ye, Z H Dong and F Yang.....87

A method of rotating accelerometer gravity gradiometer for centrifugal gradient detection

Mingbiao Yu and Tijing Cai.....94

Research on the Diesel Engine with Sliding Mode Variable Structure Theory

Zhexuan Ma, Xiaobing Mao and Le Cai.....101

Workspace Design for 6 Degree of Freedom Machining Bed

M F Shah, Z Kausar and S S Farooq.....108

Computational Model for Impact-Resisting Critical Thickness of High-Speed Machine Outer Protective Plate

Huaying Wu, Li Zhong Wang, Yantao Wang and Xiaolei Yuan.....113

Flowfield analysis of helicopter rotor in hover and forward flight based on CFD

Qinghe Zhao and Xiaodong Li.....122

Power system and energy

Probability density function evolution of power systems subject to stochastic variation of renewable energy

J Q Wei, Y C Cong and M Q Xiao.....129

Research of PV Power Generation MPPT based on GABP Neural Network

Yu Su and Xianfu Lin.....136

IOP Conf. Series: Journal of Physics: Conf. Series **1016** (2018) 011001 doi:10.1088/1742-6596/1016/1/011001

Preface

2017 6th International Conference on Mechatronics and Control Engineering took place in St. Petersburg, Russia during December 2-4, 2017. ICMCE 2017 aims to bring together researchers, scientists, developers, and engineers from around the world in both industry and academia for sharing state-of-art research results, for exploring new areas of research and development, and to discuss emerging issues in the fields of mechatronics and control engineering. It provides opportunities for the delegates to exchange new thoughts and application experiences, to establish business or research relations and to find global partners for possible future collaboration, and also it helps expanding the delegates' knowledge and insight into the significant challenges currently being addressed in the related fields.

The proceedings of ICMCE 2017 records the fully refereed papers presented at the conference, which include 22 excellent papers selected from more than 48 submitted papers whose authors are from over 13 countries. All papers published in the proceeding have been peer reviewed by the reviewers drawn from the technical committee members and external reviewers depending on the subject matter of the paper. Its scope includes but not limited to: Acoustics and Noise Control, Applied Mechatronics, Automation, Mechatronics and Robotics, Computational Fluid Dynamics, Modeling of Complex Systems, Fuzzy and Neural Systems, Sensor networks, networked control etc. After the rigorous peer-review process, the submitted papers were selected on the basis of originality, significance, and clarity for the purpose of the conference.

We would like to thank the program chairs, reviewers and the members of the program committee for their hard work. We are grateful to all those who have contributed to the success of ICMCE 2017. We hope that all participants and other interested readers benefit scientifically from the proceedings.

ICMCE 2017 Organizing Committees

December 26, 2017

IOP Conf. Series: Journal of Physics: Conf. Series **1016** (2018) 011002 doi:10.1088/1742-6596/1016/1/011002

Conference Committee

Conference Chairs

Prof. GIUSEPPE CONTE, Università Politecnica delle Marche, Italy

Prof. Magnus Thor Jonsson, University of Iceland, Iceland

Prof. Dr. Guanghsu Chang, Georgia Southern University, USA

Program Chair

Prof. Jan Awrejcewicz Biomechanics and Mechatronics, The Lodz University of Technology, Poland

Technical Committee

Assoc. Prof. Viswanathan Kodakkal Kannan, Kuwait College of Science and Technology, Kuwait

Prof. Ke-Lin Du, Concordia University, Canada

Prof. Changguo Wang, Harbin Institute of Technology,China

Assoc. Prof. Mohammad Salah, Hashemite University, Jordan

Prof. Mohamed Ichchou, Ecole Centrale de Lyon, France

Asst. Prof. Dr. Emin Taner ELMAS, Iskenderun Technical University (ISTE), Turkey

Dr. Sarunya Promkotra, Khon Kaen University, Thailand

Asst. Prof. Gokhan Erdemir, Istanbul Sabahattin Zaim University, Turkey

Dr. Zareena Kausar, Air University, Pakistan

Assoc.Prof. Mohamed Arezki MELLAL, M'Hamed Bougara University (UMBB), Algeria

Prof. Jacob Scharcanski, Universidade Federal do Rio Grande do Sul, Brazil

Asst. Prof. Mohammed N Alghamdi, Yanbu Colleges and Institutes, Yanbu Industrial College, Saudi Arabia

Assoc.Prof. Hamdy A. Ziedan, Assiut University, Egypt

Assoc.Prof. Xiaolin Wang, Engineering and Technology University of Tasmania, Australia

Dr. Binh P. Nguyen, Institute for High Performance Computing, A*STAR, Singapore

Prof. Jimoh O Pedro, University of the Witwatersrand, South Africa

Dr. Mingbiao Yu, Southeast university, China

IOP Conf. Series: Journal of Physics: Conf. Series **1016** (2018) 011003 doi:10.1088/1742-6596/1016/1/011003

Peer review statement

All papers published in this volume of *Journal of Physics: Conference Series* have been peer reviewed through processes administered by the proceedings Editors. Reviews were conducted by expert referees to the professional and scientific standards expected of a proceedings journal published by IOP Publishing.

Modeling and Sound Insulation Performance Analysis of Two Honeycomb-hole Coatings

H F Ye, M Tao and W Z Zhang

School of Mechanical Engineering, Guizhou University, Guiyang 550025, China

Email: tomn_in@163.com

Abstract. During the sound transmission loss test in the standing-wave tube, the unavoidable reflected wave from the termination of downstream tube would affect the precision measurement of sound transmission loss(TL). However, it can be solved by defining the non-reflected boundary conditions when modeling based on the finite element method. Then, the model has been validated by comparing with the analytical method. Based on the present model, the sound insulation performance of two types of honeycomb-hole coatings have been analyzed and discussed. Moreover, the changes of parameters play an important role on the sound insulation performance of honeycomb-hole coating and the negative Poisson's ratio honeycomb-hole coating has better sound insulation performance at special frequencies. Finally, it is summarized that sound insulation performance is the result of various factors that include the impedance changes, the waveform transformation and so on.

1. Introduction

The acoustic coatings that are laid on the surfaces of underwater machines and have kinds of functions such as sound absorption, sound insulation, vibration suppression, shock resistance and so on. And they are the key and the only technology that can resist active and passive detections. Acoustic coatings usually are viscoelastic composite structures that have kinds of hole types, such as spherical holes and column holes as in [1]. The coatings also can be filled with locally resonance scatterers as in [2]. While sound absorption property of these structures are good, they possess other properties as well. For example, sound insulation property of a kind of cylinder hole coating under static pressure had been analysed by Panigrahi as in [3].

Honeycomb-hole coating has been applied in many fields for its high intensity and impact resistance. As for noise control, Ruzzene has analysed the vibration and sound radiation of sandwich beams with honeycomb truss core in the air by the finite element method as in [4].And Zhao has analysed transmission coefficient and sound transmission loss of underwater honeycomb truss structures by transfer matrix method and percentage of sound energy method and got the conclusion that as for sound insulation property, this structure has better performance among circular hole rubber coating and solid rubber coating under the condition of same thickness as in [5]. As in [6], honeycomb-hole coating that is set vertically between upper and lower panels is similar to traditional cylinder hole coating in structural style, but it has bigger deformation and more influence on acoustic property under static pressure.

Changing the arrangement of the hole as the holes are set to be parallel to panel instead of the vertical arrangement has little influence on acoustic property when static pressure changes. In this paper, sound transmission property of honeycomb-hole coating that is in parallel array has been analysed. Based on the three-microphone method, sound transmission loss computational model has been established by ANSYS software; then, the influence of structural parameters on sound insulation property has been analysed; finally, the difference of sound insulation theories at low and medium high frequencies has been analysed.

2. Finite element method model and its verification

The actual measurement of TL in standing wave tube is to use the four-microphone method for the reason that complete sound absorption is difficult to realize at the end of downstream tube and unavoidable reflected wave will influence the accuracy of the result. When the FEM is adopted, the influence of the reflected wave at the end can be eliminated by modifying the end into perfect reflecting boundary condition in FEM software. So, setting one sensor in the tube is enough to measure sound pressure of transmitted wave directly.

As shown in figure 1, a finite element model for TL calculation has been established, and standing wave tube is in the form of circular waveguide and the plane wave is incident to the sample at normal incidence in the model. The side face of sample has been set as symmetry constraint to simulate the situation that the plane wave is incident to the infinite sample at normal incidence. To eliminate reflected waves at the source surface of upstream tube and the terminal surface of downstream tube and make sure that there is once incidence, once reflection and once transmission in the standing wave tube, the nonreflecting boundary condition is applied on the two faces. It should be noted that loss factor can be defined by setting the stiffness damping coefficient β . Besides, former and latter interfaces of the coating should be set as fluid solid coupling surfaces because the difference of characteristic impedance between the water and the honeycomb structure is not much.

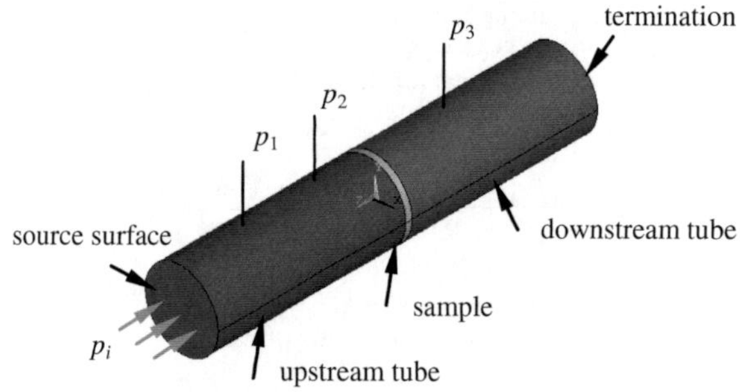

Figure 1. Finite element model for sound transmission loss calculation

In the upstream tube, the pressure p_1 and p_2 are measured independently by two microphones meanwhile. And steady state signal transmitting in the upstream tube can be resolved into two parts: one is the incident wave p_i that transmits along the positive x direction and the other is the reflected wave p_r that transmits along the negative x direction as in [7].The formula is as:

$$p_i = \frac{p_2 - p_1 e^{jkd}}{e^{-jkd} - e^{jkd}} \tag{1}$$

where d is the distance between the microphone ① and ②, p_1、p_2 are pressures measured respectively by the microphone① and②, and the pressure of transmitted wave p_t is measured by microphone③.

To verify the correctness of the model, the two-dimensional solution of acoustic characteristics of anechoic coating containing varying sectional cylindrical cavity as in [1] is used to compare. Figure 7 in [1] has showed sound absorption coefficient of the varying sectional cylindrical cavity under rigid backing condition, so only the upstream tube and the sample need to be established in the FEM model and the calculation of sound absorption coefficient is based on the transfer function method as in [8].When calculating, the pressure p_1 and p_2 are measured by microphones in the tube. This kind of comparison that sound absorption coefficients are used to compare is reasonable because calculations of sound absorption coefficient and sound transmission loss are based on the same FEM model.

Figure 2 has shown the result of comparison between the two-dimensional solution and the result of the FEM. As shown in the figure, the error between two results is relatively large at medium high frequencies because varying sectional cylindrical cavity that is the actual model had been replaced by uniform sectional cylindrical cavity that is the approximate model when analyzing based on two-dimensional model and this difference between two models made the error bigger when frequency increased. Besides, reflection coefficient of the sealing layer of cylindrical hole is 1 in two-dimensional model while the FEM model was established based on the actual shape, which would cause the error as well. Despite all this, calculation results of two methods matched well in general, which illustrates the correctness of the FEM model established in this paper.

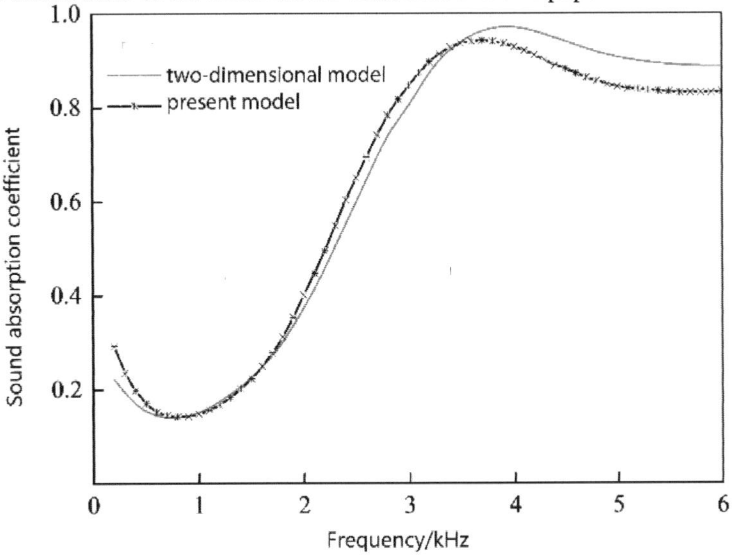

Figure 2. Comparison of sound absorption coefficient of cylindrical-hole coating

3. Analysis of sound insulation property of cylindrical-hole coating

Figure 3 shows the schematic of honeycomb-hole coating and figure 4 shows the in-plane schematic of honeycomb-hole coating. Figure 5 shows the schematic of negative Poisson's ratio honeycomb-hole coating. Two sets of parameters mainly decide sound transmission loss property of cylindrical-hole coating. One is the structural parameter shown in figure 4 as: the distance between cores of holes , $a = 5\text{mm}$;the wall thickness $t = 0.8\text{mm}$;and thickness of front and back panels is $h_f = h_b = 1.5\text{mm}$ .The other one is material parameter as: the density is $\rho = 1200\text{kg/m}^3$; Young's modulus is $E = 100\text{MPa}$;loss factor is $\eta_E = 0.2$;and Poisson's ratio is $\nu = 0.49$.

Figure 3. Schematic of honeycomb-hole coating

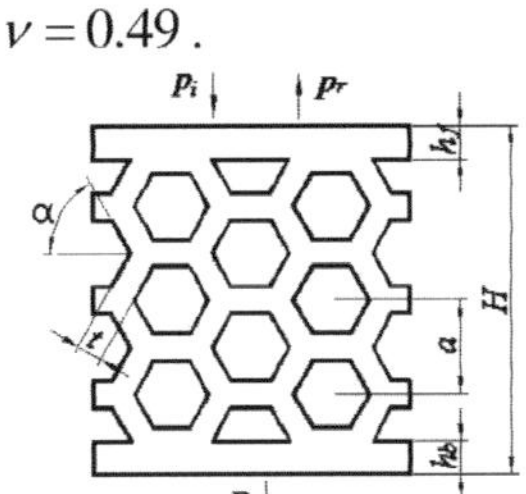

Figure 4. In-plane schematic of honeycomb-hole coating

Fig. 5. Schematic of negative Poisson's ratio honeycomb-hole coating

3.1. Influence of cell angles

Under the condition that cell thickness of honeycomb-hole coating is unchanged, the change of cell angles has influence on TL property as well. As shown in figure 4, the included angle α is the angle between the bevel edge of honeycomb and the horizontal line. TL of honeycomb-hole coating has shown in figure 6 with different angles. The change of angle has great influence on the cycle and figures of sound transmission loss. Viscoelastic waves will reflex at the angle and the change of the angle will change the impedance that will influence TL. So, the smaller angles are, the more sound transmission loss is. In other words, the smaller angles will cause the more reflection and loss, so, the transmitted energy is less and sound transmission loss is more.

Displacement contour of honeycomb-hole coating has showed in figure 7 when the angle is 45 degree and the plane wave is incident at normal incidence at 6.5kHz, which is corresponding to the frequency of the fourth maximum value in the TL curve. At the medium high frequencies, there is waveform transformation that means there is not only the longitudinal vibration but also the obvious transverse vibration in the coating, which is different from the situation at low frequency. Because the loss factor of transverse vibration is far greater than the loss factor of longitudinal vibration, TL at the medium high frequencies exceeds 30dB.

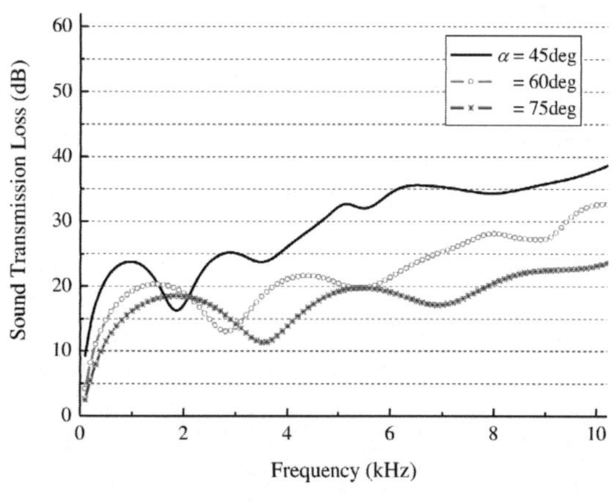

Figure 6. Sound transmission loss of honeycomb-hole with different cell angles

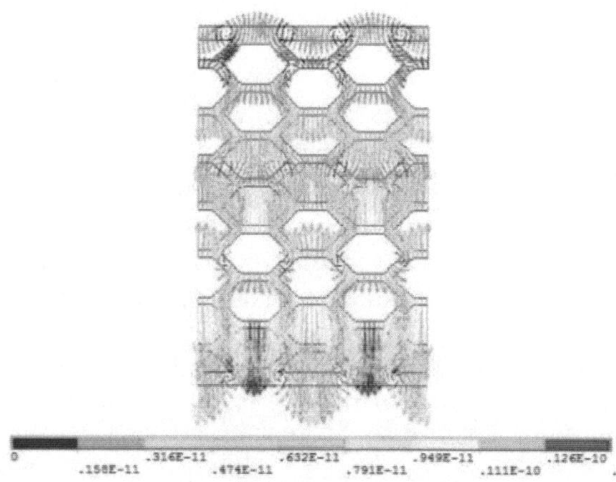

Figure 7. Displacement contour of honeycomb-hole coating,6.5KHz

3.2. Comparison with negative Poisson's ratio honeycomb-hole coating

Sound transmission loss of negative Poisson's ratio honeycomb-hole coating has shown in figure 8 with different angles that are 60°, 70° and 80° .From the result,the change of the angle has little influence on the cycle and numerical value of sound transmisstion loss at the medium high frequencies. Sound transmission loss with different angles is under 20dB when the frequencies are under 5kH, but it increases obviously when the frequencies are above 6kHz.In particular, sound transmission loss with different angles reaches the maximum value at about 7kH,8kHz and 9kHz respectively and the maximum values are over 70dB, which is different from the result of normal Poisson's ratio honeycomb-hole coating.

Displacement contour of negative Poisson's ratio honeycomb-hole coating when the angle is 70° has showed in figure 9 when the plane wave is incident at normal incidence at 7.5kHz. The main displacement happens at the excitation end of the coating. Because this kind of structure of negative Poisson's ratio can block the transmission of viscoelastic waves, displacements at the middle and back parts are not very obvious and the vibration at back part is really small. So, the negative Poisson's ratio honeycomb-hole coating has a good sound insulation property.

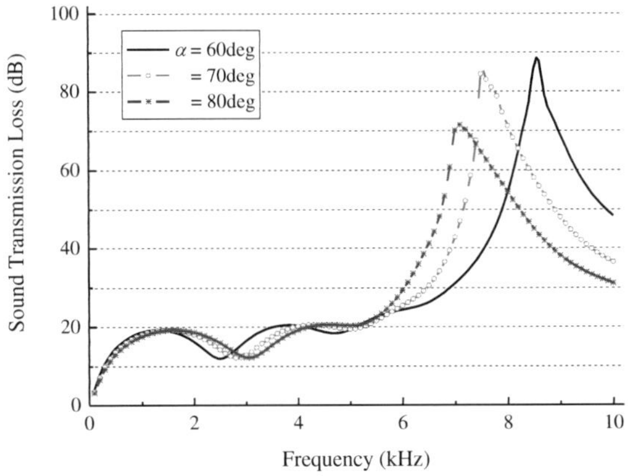

Figure 8. Sound transmission loss of negative Poisson's ratio honeycomb-hole coatings

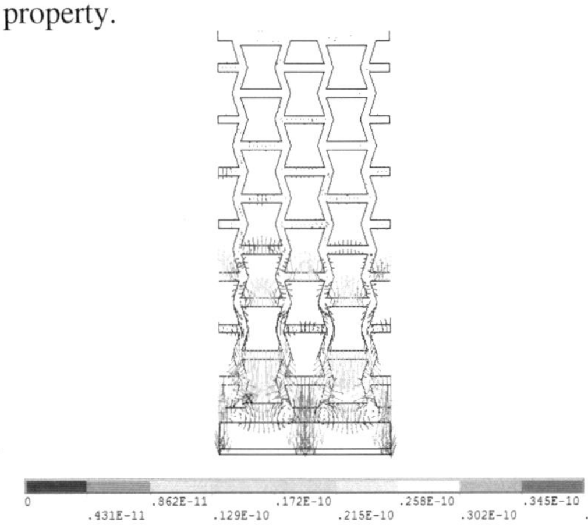

Figure 9. Displacement contour of negative Poisson's ratio honeycomb-hole coatings,7.5Khz

4. Conclusion

Based on test method of sound transmission loss in standing wave tube, a FEM model of honeycomb-hole coating has been established by ANSYS software and the correctness of the model has been validated by comparing with the result of analytic method. Then, the influence of structural parameters and material properties on sound transmission loss has been analyzed as:(1) sound insulation property of honeycomb-hole coating is influenced by multiple factors that include impedance mismatch, waveform transformation, damping loss and so on;(2) the change of cell thickness and other parameters has a prominent influence on sound transmission loss of the coating.

References

[1] He S P, Tang W L, He L, et al. 2006 *J. Ship Mech.* **10** 120-7
[2] Zhao H, Wen J, and Yang H 2014 *J. Applied Acoust.* **76** 48-51
[3] Panigrahi S N, Jog C S and Munjal M L 2008 *J. Applied Acoust.* **69** 1141-53
[4] Ruzzene M 2004 *J. of Sound and Vib.* **277** 741-63
[5] Zhao Y K, Sheng M P, Wang C J, et al. 2006 *J. Mech. Sci. Tech.* **25** 745-7
[6] Tao M and Zhuo L K 2011 *J. of Shanghai Jiaotong University* **45** 1340-4
[7] Zhu B L and Luo X H 2000 *J. Noise Vib. Control* **20** 41-3
[8] Zhu B L and Xiao J X 1994 *J. Acta Acoust.* **19** 351-60

Acknowledgements

This work was supported by the Natural Science Foundation of China Grant No. 51765008, Guizhou High-level Innovative Talents Program Grant No.2016-4033, and Guizhou University Innovation and Entrepreneurship Training Program Grant No. 2014-010.

IOP Conf. Series: Journal of Physics: Conf. Series **1016** (2018) 012002 doi:10.1088/1742-6596/1016/1/012002

Composite adhesive bonds reinforced with microparticle filler based on egg shell waste

Miroslav Müller [1], Petr Valášek[1]

[1]Czech University of Life Sciences Prague, Faculty of Engineering, Department of Material Science and Manufacturing Technology, Kamýcká 129, Prague, Czech Republic

E-mail: muller@tf.czu.cz

Abstract. A research on composite adhesive bonds reinforced with waste from hen eggs processing, i.e. egg shell waste (ESW) is based on an assumption of the utilization of agricultural/food production waste. The aim of the research is to gain new pieces of knowledge about the material utilization of ESW, i.e. to evaluate possibilities of the use of various concentrations of ESW microparticles smaller than 100 μm based on hen egg shells as the filler in a structural resin used for a creation of adhesive bonds from bearing metal elements. An adhesive bond strength, an elongation at break and a fracture surface were evaluated within the research on adhesive bonds. The experiment results proved the efficiency of ESW filler in the area of composite adhesive bonds. The adhesive bond strength was increased up of more than 17 % by adding 40 wt.% of ESW microparticles.

1. Introduction

The research focusing on the utilization of fillers based on renewable resources is essential. Renewable resources are a suitable material for a substitution of synthetic materials /fillers [1, 2, 3, 4, 5]. The production of waste where also secondary products /waste can be ranked in many cases is a worldwide topic – many secondary materials are burnt or otherwise liquidated without a possibility of alternative handling [1, 3]. The utilization of natural reinforcement leads to a reduction of a mass and a prices of single parts and to an optimization of mechanical properties [1, 2, 3]. A meaningful utilization of secondary materials /waste as the material utilization is an economic and environmental alternative to other handling possibilities or to a liquidation [1, 2, 3, 6, 7, 8]. The research used to be focused on particle and fibre composite materials above all. An area of structural adhesive bonds is one of prospective possibilities of a practical utilization of particle nano and micro filler, fibres in composite materials [9, 10, 11].

The research deals with the utilization of a by-product from the food industry, namely waste coming into being at processing hen eggs. The production of hen eggs on an industrial level leads to an essential amount of shells which are regarded as the waste or they are used as a supplement in the agriculture [12]. Generally it can be said that the egg shells are of no economic value [13]. However, they are rich of minerals. This material can be regarded as a source of the calcium filler [14]. Many studies proved their positive influence as an ingredient in various products [13]. It has been demonstrated that using low-cost materials and a relatively simple process complex supports with a mixed inorganic–organic structure can be synthesized [15].

About several hundred thousand tons of egg shell waste come into being at the industrial processing of eggs. Egg shells represent ca. 11 % of total mass of eggs and they are available as the waste in huge amounts namely in food industry [13]. Egg shell waste is potentially suitable candidate for an ecological filler material for reinforced biopolymeric composites owing to improving their mechanical properties and a heat stability [13]. E.g. Zieleneiwska et al. dealt with the composite materials based on polyurethane foam with egg shells used as the natural filler [13]. They dealt with a determination of the filler content on a structure of the material and mechanical properties.

The aim of the research is to gain new pieces of knowledge about material utilization of ESW, i.e. to evaluate possibilities of the use of various concentrations of ESW microparticles smaller than 100 μm based on hen egg shells as the filler in the structural resin used for creating adhesive bonds from bearing metal elements. The paper deals with the utilization of a by-product from the food industry, namely waste coming into being at processing hen eggs, i.e. egg shells waste effectively usable as the microparticle filler in the area of composite adhesive bonds. An adhesive bond strength, an elongation at break and a fracture surface were evaluated within the research on adhesive bonds. A scanning electron microscopy (SEM) was used for a description of an interaction of ESW filler and a resin.

2. Materials and Methods

Microparticles in a form of egg shell waste (ESW) were used within the research. ESW was obtained from the food industry and dried for the time 24 h at the temperature 105 ± 5 °C. Subsequently, ESW was grinded in an industrial grinder. A fractionalisation was performed on sieves with a mesh size 100 μm. The particle analysis was performed at the device Haver EML digital plus. The fraction with the smallest particles, i.e. in the interval 0 to 100 μm was used for the research. The egg shell microparticles were dried before the application to the resin at the temperature 105°C for the time 24 h. Composite adhesive bonds were the subject of performed experiments. The structural carbon steel S235J0 was the adhesive bonded material. A layer of the adhesive mixture, i.e. the adhesive was composite, i.e. the continuous phase was in a form of structural two-component epoxy resin CHS EPOXY 324 epoxy 1200 (a matrix / resin – marked as 0 wt.%). Microparticles of ESW were the discontinuous phase (reinforcing particles).

Adhesive bonds were prepared in accordance with the standard ČSN EN 1465, i.e. laboratory tests were performed using the standardized test specimens of dimensions 100 ± 0.25 x 25 ± 0.25 x 1.5 ± 0.1 mm and lapped length of 12.5 ± 0.25 mm. The adhesive bonded surface of the carbon steel was mechanically treated – grit blasted by Garnet MESH 80 and chemically treated – cleaned in the acetone bath. Roughness parameters were measured with a portable profilometer Mitutoyo Surftest 301. A limit wavelength of the cut-off was set as 0.8 mm. The surface roughness at the grit blasted adhesive bonded material, i.e. the structural carbon steel S235J0 was Ra = 1.75 ± 0.19 μ m, Rz 11.42 ± 0.87 μ m.

The composite adhesive was prepared by adding the filler based on ESW microparticles in 10, 20, 30 and 40 wt.%. The matrix is mid-molecular resin consisted of bisphenol A and epichlorhydrin (an average molecular mass ≤ 700). P11 was the hardener containing amines, polyethylenpoly, triethylentetramine fraction. The composite adhesive was prepared by mixing of the resin and the filler in the form of ESW microparticles for the time 10 min. This step was performed owing to the short time of the adhesive mixture processing and a necessity of a sufficient homogenisation of prepared mixture of the filler and the resin. Subsequently, the hardener P 11 was added.
Subsequently, the composite adhesive was deposited on the prepared surface of the structural carbon steel S235J0. Adhesive bonds were hardened for 72 ± 5 h with a temperature 22 ± 2°C. Adhesive bonds were fixed with the weight of 750 g.

Mechanical characteristics of adhesive bonds based on the composite adhesive were performed by means of the universal tensile strength testing machine LABTest 5.50ST (a sensing unit AST type KAF 50 kN, an evaluating software Test&Motion). A loading speed of a deformation corresponded to 5 mm.min-1.

A scanning electron microscopy (SEM) was used for a description of the interaction of ESW filler and the resin. The measurement was performed on the scanning electron microscope TESCAN MIRA 3 GMX. SE and BSE detector were used for the measurement, the accelerating voltage was 5 kV and the working distance was ca. 15 mm. The samples were dusted with gold by means of the equipment

IOP Conf. Series: Journal of Physics: Conf. Series **1016** (2018) 012002 doi:10.1088/1742-6596/1016/1/012002

Quorum Q150R ES - Sputtering Deposition Rate using Gold. SEM analysis helps to understand results of mechanical tests. Measured values were processed by means of statistical analysis methods. ANOVA test was used for the statistical comparison of measured data.

3. Results and Discussion

Results of the adhesive bond strength are visible in fig. 1 A. It is obvious from the results that the adhesive bond strength was increased by adding ESW microparticle filler. The adhesive bond strength was gradually increased with increasing wt.% concentration of ESW microparticle filler, namely up of 17.84 % Higher concentration than 40 wt.% was not possible to create. The structural epoxy resin was saturated with the hardener P11. It is possible to say in terms of the statistical testing of the influence of ESW microparticle concentration by means of ANOVA F-test in the significance level 0.05 that the concentrations are statistically non-homogeneous groups ($p = 0.0001$), i.e. there is the difference in the resultant adhesive bond strength among single tested concentrations. It is obvious from the results that different concentrations influence the adhesive bond strength.

Results of the elongation at break of adhesive bonds are visible in fig. 1 B. It is obvious from the results that the elongation at break of the adhesive bond did not change by adding the ESW microparticle filler. The elongation at break of the adhesive bond was increased with increasing wt.% concentration of ESW, namely up of 13.3 %, it is true, but a dispersion of results was increased from 8.5 % (the resin – 0 wt.% filler) to the interval from 8 to 17 % depending on the concentration of ESW microparticle filler. It is possible to say in terms of the statistical testing of the influence of ESW microparticle filler by means of ANOVA F-test that the concentrations are statistically homogeneous groups ($p = 0.0777$) in the significance level 0.05, i.e. there is no difference in the resultant elongation at break of adhesive bonds among single tested concentrations. It is obvious from the results that different concentrations do not influence the elongation at break of the adhesive bond.

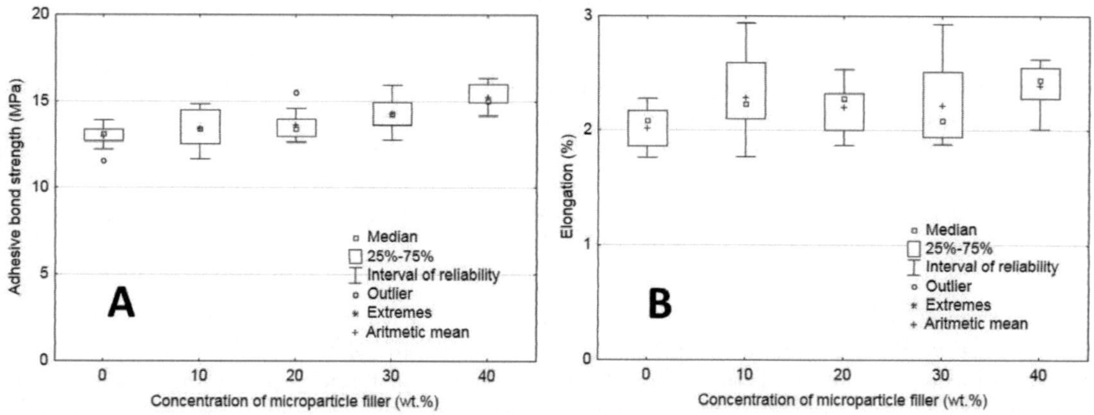

Figure 1. Results of experiment: A: Influence of concentrations of ESW microparticle filler on adhesive bond strength, B: Influence of concentrations of ESW microparticle filler on elongation at break of adhesive bond.

A fracture surface of the matrix (the resin – 0 wt.% filler) and the composite bond with 10 wt.% of ESW microfiller was of an adhesive type, i.e. it came to a destruction in the interface of the layer of the adhesive and the adhesive bonded material. The fracture surface changed to the adhesive / cohesive by adding ESW microparticles of the concentration 20 to 40 wt.%. Results of SEM analysis proved the change of the fracture surface (fig. 2). It is obvious from fig. 2 B different distribution of ESW microparticles in the constant layer of the adhesive. A detailed view on the interaction between the microparticle filler based on the ESW and the resin is visible from fig. 2. A good wettability is visible from fig. 2 B, C in the layer of the adhesive. SEM analysis proved a good dispersion of the filler in the matrix (fig. 2). Also Krishnan et al. came to similar conclusions [16]. A shape of ESW microparticles is visible from fig. 2. It is evident from fig. 2 that ESW is of very porous structure. A bad wettability and a non-homogeneity of the natural fillers utilized in the polymeric composites belong among main disadvantages of their use [2, 8]. However, this was not proved.

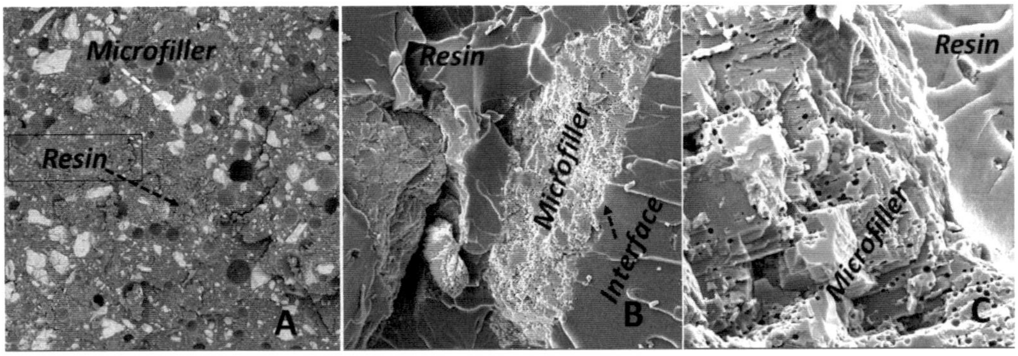

Figure 2. SEM images of fracture surface of adhesive layer: A: distribution of ESW microparticles – composite layer of adhesive - 30 wt.% ESW (MAG 267 x), B: composite layer of adhesive - 40 wt.% ESW (MAG 4.64 kx), C: composite layer of adhesive - 40 wt.% ESW – detailed view on porous structure of microparticle (MAG 15.8 kx).

Egg shells contain about 95% of calcium carbonate (calcite) and 5% of organic materials such as sulphated polysaccharides, type X collagen and other proteins [13]. Filling of thermoplastics by the calcium carbonate is used in the practice when this filler improves mechanical properties [12]. A positive effect of the filler based on egg shell microparticles in the area of the composite materials is visible from the research results of Zieleniewska et al. Also results of there experiments confirmed this fact [13]. The small particle size aids in a uniform stress transfer between the filler and the matrix, thereby resulting in a composite with superior mechanical performance [17].

4. Summary

The paper described a possible utilization of ESW microparticles, i.e. the secondary raw materials from eggs processing in the food industry, in the area of the composite adhesive bonds. It is desirable to incorporate renewable resources or waste from animal production in the material engineering. The experiment results proved the efficiency of ESW filler in the area of the composite adhesive bonds. Following conclusions can be stated from the results of the experiment using the natural filler based on ESW microparticles:

- The adhesive bond strength is increasing with increasing concentration of ESW microparticles.
- The elongation at break of the adhesive bond does not significantly change with adding ESW microparticles.
- ESW microparticle filler proved good wettability with the matrix, i.e. the resin based on SEM analysis.

5. References

[1] Ruggiero A, Valášek P and Müller M 2016 *Compos Part B – Eng.* **104** pp 9–16
[2] Mizera Č, Herák D, Hrabě P, Müller M and Kabutey A 2017 *Journal of Natural Fibers* **14** pp 287–296
[3] Müller M, Valášek P and Ruggiero A 2017 *BioResources* **12** pp 255–269
[4] Amuthakkannan P, Manikandan V, Winowlin Jappes J T and Uthayakumar M 2013 *Materials Physics and Mechanics* **16** pp 107–117
[5] Faruk O, Bledzki A K, Fink H P and Sain M 2012 *Progress in Polymer Science* **37** pp 1552–1596
[6] Valášek P, Ruggiero A and Müller M 2017 Compos Part B: Eng. **122** pp 79–88
[7] Valášek P and Müller M 2015 Tehnicki Vjesnik **22** pp 257–262

[8] Herrera-Franco P J and Valadez-Gonzalez A 2005 *Compos: Part B. Eng.* **36** pp 597–608
[9] Müller M, Valášek P and Rudawska A 2017 *Journal of Adhesion Science and* Technology **31** pp 1859–1871
[10] Müller M 2015 *Agronomy Researc*h 13 pp 700–708
[11] Müller M, D´Armato R and Rudawska A 2017 *16th International Scientific Conference Engineering for Rural Development*. Jelgava 24-25.6.2017 Latvia pp 121–127
[12] Petrásek P and Müller M 2017 *16th International Scientific Conference Engineering for Rural Development*. Jelgava 24-25.6.2017 Latvia pp 175–180
[13] Zieleniewska M, Leszczynski K M, Szczepkowski L, Bryskiewicz A, Krzyzowska M, Bien K and Ryszkowska J 2016 *Polymer Degradation and Stability* **132** pp 78–86
[14] Oliveira D A, Benelli P and Amante E R 2013 *Journal of Cleaner Production* **46** pp 42–47
[15] Badanoa J M, Betti C, Rintoulb I, Vich-Berlanga J, Cagnola E, Torres G, Vera C, Yori J and Quiroga M 2010 *Applied Catalysis A: General*. **390** pp 166–174
[16] Krishnan A, Iyer A and Torkelson JM 2014 *Composites Science and Technology* **102** pp 152–160
[17] Cheung H Y, Ho M P, Lau K T, Cardona F and Hui D 2009 *Composites: Part B*. **40** pp 655-663

Acknowledgments
Supported by Internal grant agency of Faculty of Engineering, Czech University of Life Sciences Prague no. 2017:31140/1312/313115.

IOP Conf. Series: Journal of Physics: Conf. Series **1016** (2018) 012003 doi:10.1088/1742-6596/1016/1/012003

Degradation of strength properties of epoxy resin filled with natural-based particles

Petr Valášek[1], Karolína Habrová[1], Miroslav Müller [1]

[1]Czech University of Life Sciences Prague, Faculty of Engineering, Department of Material Science and Manufacturing Technology, Kamýcká 129, Prague, Czech Republic

E-mail: valasekp@tf.czu.cz

Abstract. Degradation of polymeric materials can be considered as a limiting factor for their use. Mechanical characteristics of epoxy resins are reduced, for example, by the action and changes of temperature or humidity. Degradation also occurs in composite systems where the epoxy resins function as matrices, i.e. in polymer composite materials. If a natural filler is used together with the epoxy resin, we refer to these materials as biocomposites, where also the natural character of the filler material greatly affects the degradation process. The paper focuses on the description of the shear strength of the resin filled with particles prepared from the seeds of dates of Phoenix Dactylifera plant. The degradation was evaluated experimentally in laboratory conditions via the climatic chamber. The experiment describes composites with a particle size of filler 100-200 μm with a concentration of 5 - 10 wt%. As the number of degradation cycles increased, the tensile strength of both the unfilled and the filled epoxy resin decreased. After 5 weeks, the drop was up to 50%. The presence of the particles did not significantly affect the shear strength compared to the non-filled resin. The described way of utilization of the natural-based particles is the possibility of material utilization of secondary natural materials.

1. Introduction

At present, in the area of material engineering, the emphasis is placed on saving primary resources that can be readily replaced by secondary resources. The possibility of this substitution is filled with natural materials. A very promising area is the possibility of using secondary products on a natural basis in composite systems. [1, 2]

Natural materials can be in form of both particle and fiber in the composites. Particles are mainly used to optimize abrasive wear, they can increase shear strength or cohesive characteristics of the resulting material [3, 4]. Factors affecting the resulting properties of the composite system include particle size, morphology, surface properties and, last but not least, their concentration in the material. [5-7]

Secondary raw materials on a natural basis can substitute primary sources. Their meaningful use is also desirable from an economic point of view. Such materials can be considered, for example, as seeds that are produced during the processing of Phoenix Dactylifera plants, that is, dates [8]. Said et al. [9] describe that there are many scraps of Phoenix Dactylifera, which is produced on such a secondary basis - for example, in Algeria it refers to 85,000 tons/year. Such secondary products can be used energetically, another possibility is their material utilization where they can substitute materials

IOP Conf. Series: Journal of Physics: Conf. Series **1016** (2018) 012003 doi:10.1088/1742-6596/1016/1/012003

such as wood - wood flour, which is used in the field of material engineering to reduce the price of material while maintaining the required mechanical characteristics. An example of a similar substitution may be, for example, the work of Crespo et al. [10] who uses secondary products in composite systems in the form of almond husks, obtained as a by-product of the agri-food industry. The use of Phoenix Dactylifer particles in composite systems is described, for example, by Al Maadeed [11]. Polymeric materials, including epoxy resins, that are exposed to a specific environment for a certain period of time, somehow change their properties - they degrade. Degradation processes are caused by a variety of factors, including temperature, humidity, chemical effects, weather effects, and more. Materials, also polymers, lose their required properties by acting of named factors, which is reflected, for example, by a decrease in mechanical characteristics.

The aim of the experiment was to assess whether particles based on secondary renewable raw materials prepared from date seeds affect the shear strength of epoxy resin on steel adherents, and also to assess the degradation process of these composite systems in comparison with the unfilled resin.

2. Materials and Methods

Seeds of the dates can be considered as a secondary commodity resulting from the processing of Phoenix Dactylifera plants. Dates can be considered as a fruit that has been used for many years in a row and its use is still up to date. Effective use of all commodities arising during dates processing is important from the economic point of view of the entire process.

The seeds of dates grown in China were crushed with a knife mill at a speed of 20,000 rpm·min-1, dried at temperature 105 ° C and subsequently fractionated by analytical sieves. The fraction was formed between sieves of 100 and 200 μ m.

The two-component epoxy resin was used as a matrix. This resin evinces excellent adhesion to a variety of materials and is therefore suitable for use in composite systems. The prepared microparticles were mechanically mixed with the epoxy resin to obtain the desired concentration of these particles in the composite system, namely 5 and 10 weight percent (wt%). The resulting composite mixture was applied to conventional carbon steel sheets (S235J0). These sheets were blasted before application of the composite mixture and their roughness was assessed by roughness parameters Ra and Rz, steel sheets were also cleaned and degreased prior to application of the composite composition. The degradation of the specimens was carried out in the climatic chamber according to the standard CSN EN ISO 9142, where the particular cycles were changed. Each cycle consisted of individual parts, see Tab. 1. There was always a rapid change of the desired parameters between the different parts of the cycle, within 15 minutes. These time periods are counted in the total cycle time, which corresponds to 24 h. Total 35 cycles (840 h) were performed, corresponding to 5 weeks. Samples were removed from the temperature chamber every 7 cycles - 1 week.

Table 1. Description of the degradation cycle in the climatic chamber.

Parametr	1 part	2 part	3 part
Temperature [°C]	70	-40	70
Humidity[%]	90	-	50
Time [h]	16	3	5

The adhesive strength of the composite system on the steel adherent was evaluated according to CSN EN 1465 after removal from the degradation chamber - by defining the shear strength of these joints. After this destructive test, the types of disruption were assessed.

Electron microscopy (Tescan Mira 3 GXM) was used to describe morphology and particle size and was used to describe the interfacial interaction between the matrix and microparticle particles prepared from date seeds and to evaluate the fracture areas after a destructive test. To verify the hypothesis whether the presence of particles prepared from date seeds affect shear strength compared to a non-filled resin and whether the degradation process affects the resulting shear strength of the composite system on steel adherence, ANOVA and T-test were used. The truth of the hypothesis H0 that the given factor does not affect the observed quantity indicates parameter p> 0.05.

3. Results and Discussion

Description of topography and particle size are crucial to the resulting properties of composite systems. The shape and size of the particles influence both the cohesive and adhesive characteristics of the composite system. Particle based on seeds of dates have an irregular shape that supports mechanical interfacial bonding. The particle size is indicated by the following Tab. 2. The mean particle size was 148.83 ± 28.87 μ m, ranging from 67 μ m to 197 μ m.

Table 2. Description of particle sizes prepared from date seeds - frequency of observation.

Particle size [μm]	<100	100-119	120-139	140-159	160-179	180-200	200>
Observations [%]	7	9	16	31	25	12	0

Prior to the degradation evaluation, the strength of the unfilled epoxy resin (12.28 ± 0.69 MPa) was compared with the system filled with 5 wt% (12.37 ± 0.60 MPa) and 10 wt% (12.36 ± 0.83 MPa). At the significance level $\alpha = 0.05$ it was shown that there is no statistically significant difference between unfulfilled resin and composite systems where the parameter for the composite 5 wt% was p = 0.82, and the H_0 hypothesis for 10 wt% was p = 0.86. Roughness parameters and surface cleanliness are among the basic factors influencing the adhesion strength between the composite system formed by the matrix of epoxy resin and the used adherent. Optimizing these factors is crucial for interfacial interaction on this interface. The basic parameters of the roughness of the blasted steel sheet prior to the application of the composite system corresponded to Ra = 2.31 μm and Rz = 11.93 μm.

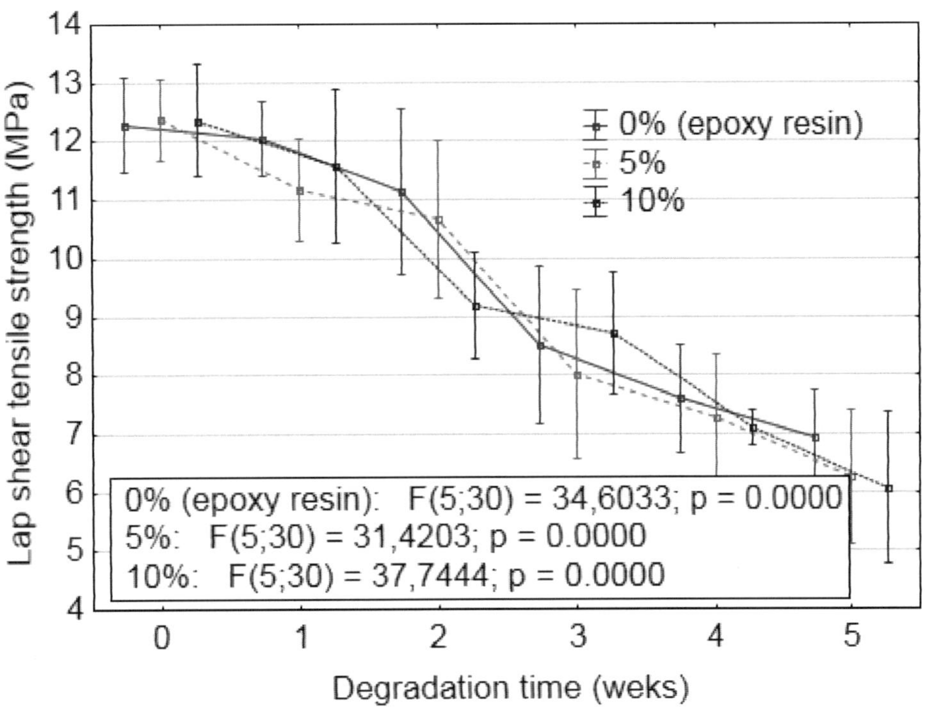

Figure 1. The dependence of the values of shear strength of filled and unfilled resin on degradation time.

In the case of a non-filled resin, the shear strength dropped by 5.36 MPa. In the composite system with 5 wt% of particles caused a decrease of 6.12 MPa and 10 wt% in the system reached 6.29 MPa. The lowest shear strength after 5 weeks of exposure was recorded in the composite system with 10 wt% of the filler - value 6.07 ± 0.98 MPa. Statistical comparison of the degradation process, i.e. comparison with a data set that was not subject to degradation at a significance level $\alpha = 0.05$, is shown in the following Table. 3.

Table 3. T-test: Effect of degradation on shear strength - parameter p.

H_0: $\mu_1=\mu_2$ (p>0.05)	**Degradation (weeks)**				
	1	**2**	**3**	**3**	**5**
0 wt%	**0.57**	**0.10**	0.00	0.00	0.00
5 wt%	0.01	0.01	0.00	0.00	0.00
10 wt%	**0.24**	0.00	0.02	0.00	0.00

After the first week of degradation, the decrease in shear strength values was statistically demonstrated in the resin filled with 5 wt% of the particles. After the second week of degradation, a statistically demonstrable decrease in shear strength was observed in both composite systems. In the case of non-filled resin, the decrease was only apparent after 3 weeks of degradation.

The interfacial interaction was evaluated by electron microscopy, on fracture surfaces after the destructive tests, see Fig. 2. The nature of the infringement was in most cases an adhesive type of failure. Failure thus occurred at the interface of filled resin and steel adherent. An increase in interaction at this interface could be achieved, for example, by another treatment of steel adherents prior to application of the resin, but blasting and subsequent surface degreasing are common treatments of materials where the epoxy resin is applied. [4, 5]

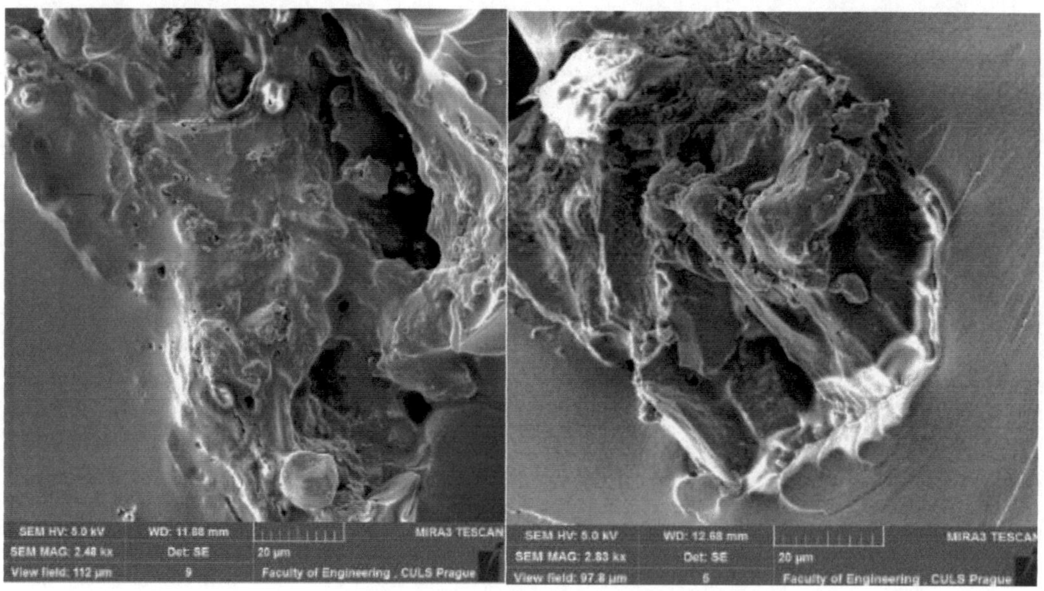

Figure 2. Interfacial interaction assessment – SEM: Mag 2.48 kx (left), Mag. 2.83 kx (right).

From Fig. 2 is apparent, that there is a good interfacial interaction between used particulate filler and the epoxy resin. For example, chemical surface treatment of particles could increase this interaction with natural fillers [2]. The experiment showed that natural particles create good interactions with epoxy resins that lead to the formation of composite systems with the desired properties - there is no reduction in shear strength - this assumption confirms a number of works [1, 9, 11]. The described way of using natural secondary materials is environmentally sensitive. Particles prepared from the Phoenix Dactylifera plant by processing the seeds, for example, can substitute the use made of a wood flour.

4. Summary
The effect of temperature changes in the range of -40 ° C to + 70 ° C, along with changes in the humidity of 50 to 90%, significantly affected shear strength. This exposure of the test specimens in the degradation chamber for 5 weeks resulted in a decrease in the strength of the non-filled resin by 43.6%. For composite systems filled with microparticles prepared from date seeds, this drop was 49.5% (5 wt%) and 50.9%. (10 wt%). A slight increase in the rate of the degradation of shear strength was observed as a result of the presence of a particulate filler from the date processing. However, the natural-based microparticle filler does not significantly affect the shear strength of the resin prior to degradation.

5. References
[1] Fowler P A, Hughes J M and Elias R M 2006 *J. of the Science of Food and Agriculture* **86** pp 1781–1789
[2] Valášek P, Ruggiero A and Müller M 2017 *Comp. Part B – Eng.* 122 pp 79–88
[3] Ruggiero A, Merola M, Carlone P, Archodoulaki V M 2015 *Compos Part B: Eng*, 79 pp 595–603
[4] Soykok I F 2017 *Journal of Adhesion* **93** pp 357–374
[5] Müller M, Valášek P and Ruggiero A 2017 *BioResources* **12** pp 255–269
[6] Valášek P and Müller M 2015 *Tehnicki Vjesnik* **22** pp 257–262
[7] Valášek P 2014 *Manufacturing Technology* **14** pp 632–637
[8] Ruggiero A, Valášek P and Müller M 2016 *Compos Part B – Eng.* **104** pp 9–16
[9] Said A, Leila D and Kaouther, B. 2014 *Sadia:Iranian Journal of Biotechnology* **12** pp 1006–1015
[10] Crespo J E, Sanchez L, Parres F and López J 2007 *Polymer Composites* **28** pp 71–77
[11] Al Maadeed M A, Kahraman, R, Noorunnisa Khanam P and Al-Maadeed S 2013 *Materials and Design* **43** pp 526–531

Acknowledgments
Supported by Internal grant agency of Faculty of Engineering, Czech University of Life Sciences Prague no. 2017:31140/1312/313113.

IOP Conf. Series: Journal of Physics: Conf. Series **1016** (2018) 012004 doi:10.1088/1742-6596/1016/1/012004

Experience of light thin-walled structures improvement in construction

A V Frolovskaia[1], S V Deordiev[1], A Falk[2], N Y Klinduh[1] and I I Terehova[1]

[1] School of Engineering and Construction, Siberian Federal University, 82 Svobodny, Krasnoyarsk, 660041 Russia
[2] KTH Royal Institute of Technology, SE – 10044, Stockholm, Sweden

E-mail: tereshkova81@mail.ru

Abstract. The authors on the basis of practical experience have analyzed low-rise construction with the use of energy-saving technologies. Characteristic features of possible variants of frame construction are looked at and described. The relevance of the paper consists in the improvement of the building frame design solution based on the analysis and elimination of disadvantages taking into account consumers' point of view.

1. Introduction

At the present time low-rise construction takes on the appearance of mass character and enjoys specific popularity; it refers to the region where innovative energy saving technologies and materials are being introduced most rapidly and effectively. High cost of building materials, lack of knowledge, absence of clear requirements and experience in the sphere of energy efficient design make the realization of most projects unprofitable.

One of the most popular technologies of construction of low-rise buildings is a frame building of light steel thin-walled structures (LSTWS) [1, 2, 3, 4].Among its advantages manufacturers indicate the resistance to corrosion of galvanized profiles, low weight, lack of shrinkage, quick installation; wall structure do not require preparatory work for finishing. At the same time among the deficiencies they indicate only the suitability of the technology in low-rise construction and unsuitability for the buildings located on heaving soils. The experience of the use and the LSTWS basic technology prerequisites are described in [5, 6, 7].

This article presents the practical experience of construction from thin-walled cold-formed profiles. The authors' task is to share the experience of improvement of constructive solutions on the basis of the analysis of shortcomings and common errors in frame-house construction, including their elimination relying on the opinions of direct consumers as well.

2. Application of LSTWS at construction of apartment houses

The particular feature of traditional constructive solution is a frame of the building, consisting of thin-walled steel struts (S), connected at the top and bottom by the strapping beams (SB1, SB 2) and additional links (L1). Spatial resistance and the rigidity of the frame on a whole is provided by the formed system of longitudinal and transverse load-bearing walls, rigidly connected by a floor structure. Layout of thin-walled galvanized cold-formed profiles in plan and in section is shown in Figure 1.

IOP Conf. Series: Journal of Physics: Conf. Series **1016** (2018) 012004 doi:10.1088/1742-6596/1016/1/012004

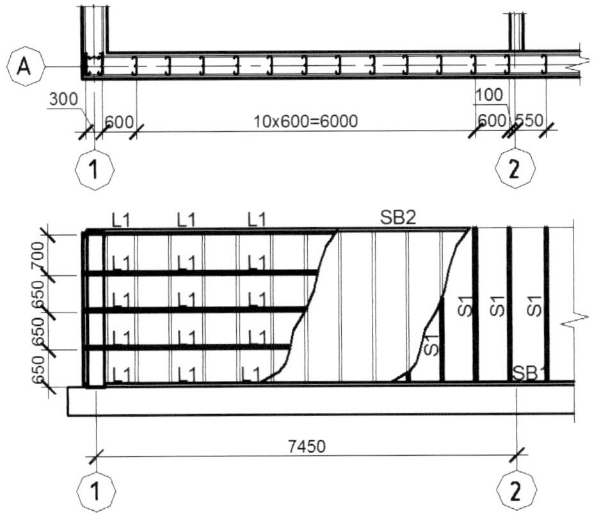

Figure 1. Layout of the thin-walled galvanized profiles in plan and in section

This technology was used in the construction of single-storey semi-detached residential houses built in accordance with "the Program of resettlement from dilapidated housing" in the Birilyuss district of the Krasnoyarsk territory (Figure 2). At the core of the frame C-shaped stands with 50x150x50 mm section of 1.5 mm thick are provided, arranged with a gap of 600 mm with stacked plates of the mineral wool insulation of 150 mm and 50 mm thick. To ensure the frame spatial rigidity the stands are tied with vertical Z-shaped and U-shaped inclined ties. To exclude the "cold bridge" between the steel elements, strips from waterproof plywood are provided. On the inner side walls are sheathed with gypsum fiber plates in 2 layers; between them the vapor barrier is provided with siding from the outside. As bearing elements, cover trusses are provided supported directly on the rack frame. To ensure the stability of frames, their upper belt is fixed by Π-shaped spans with a step equal to the grid coverage element step. To reduce the consumption of materials, the insulation coating was carried out along the horizontal plane at the level of the covering plate.

Figure 2. Housing frame under construction

The application of the above technology in the construction of houses is justified not only by the forced short construction time, but also by the need to use a cost-effective option. The area of one building is 156.6 m^2. Technical and economic indicators of the project are the following: cost of construction – 53,330 $, duration - 3 months, the cost of 1 m^2 is the 340.5 $ including delivery of materials to the construction site at a distance of 250 km.

The calculation of the building frame was made according to the spatial pattern with the reduction of the cross-sections of elements. In the construction of houses the respective authors' and quality control were provided. Residents moved into their homes more than 2 years ago with no complaints so far.

3. Improvement of the LSTWC technology in the case of an administrative building

Taking into account the opinion of consumers who have used the LSTWS technology in low-rise construction, we can conclude that there is a large number of projects with defects caused not only by saving on good quality construction materials and installation time, but also by the development of design-estimate documentation. After analyzing the information from various forums [8, 9], where the opinion is expressed not only by manufacturers and builders, but also by the consumers themselves one can make a conclusion that the population has a bias towards this technology.

Most of the buildings built with the LSTWS technology, are not able to find their buyer. People refuse to buy apartments in the houses, considering that the apartments cannot be used in the form of comfortable housing. The main disadvantages that the residents identify are: lack of competent approach to the creation of the project, including its construction; the hollowness of the structures of walls and ceilings [8]; the lack of information on the effects of electromagnetic radiation for safe living in buildings with a metal frame [10]. One of the main issues for future residents is a problem of furniture mounting to the ''thin" walls, which is solved by additional strengthening of cross platforms from boards and beams. During the period of construction it is impossible to predict where the furniture will be located. Afterwards, with the introduction of the building the arrangement of additional elements reduces the useful area of the house and requires additional work

The results of consumers ' opinions were taken into account in the construction of low-rise administrative building with the use of the LSTWS technology.

The technology which received a positive feedback during the construction of one-storey houses was used as the basis of constructive solution. The customer suggested to cut metal consumption by reducing the cross section of load-bearing profiles (50x100x50 mm), and by binding racks on two sides by braced profiles and wooden bars (50-50 mm with a step of 600 mm) to provide insulation in 3 layers – Figure 3. The racks step was 600 mm depending on the size of the insulation slab.

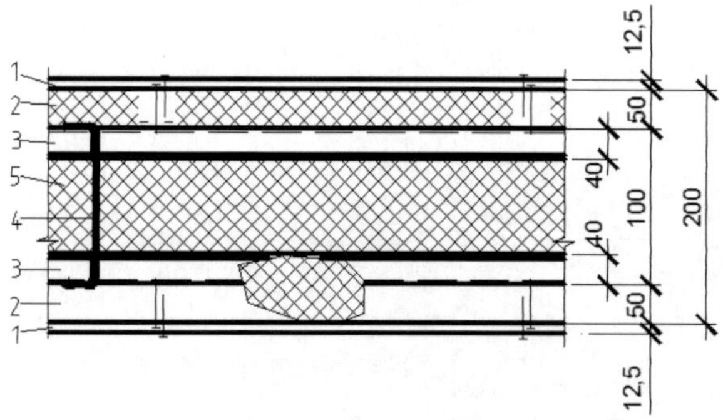

Figure 3. Constructive solution of the wall (variant 1):
1. Gypsum – fiber board
2. 50x50mm wooden block
3. Binding profiles
4. Frame bearing rack
5. Mineral wool insulation

As an alternative of the first variant, the authors of the project proposed to address the technology, when steel thin-walled profiles play the role of a bearing skeleton of the building only during the construction period. At the same time, these elements serve as guides for the formwork fastening - Figure 4. When performing calculations of the bearing capacity of the frame elements, loads acting during the building maintenance are adopted. By reducing the cross-section of the elements the coefficient of their utilization was close to one and the bearing elements of the frame did not have sufficient margin of safety.

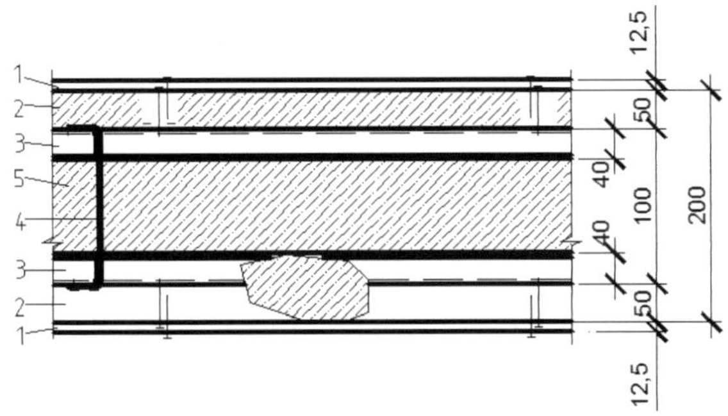

Figure 4. Constructive solution of the wall (variant 2):

1. Gypsum – fiber board
2. 50x50mm wooden block
3. Binding profiles
4. Frame bearing rack
5. Monolithic foam concrete

Figure 5 shows a schematic arrangement of bearing elements of the frame in plan and in section of the improved LSTWS technology. The calculation of the frame is made in two stages: during installation and during operation. The result is that the bearing capacity of all the elements is provided, the utilization coefficient of the cross section equals to 0.8. Mounting values are taken as the loads on the steel frame of the building. During the period of operation of the building, both the steel frame and monolithic foam concrete are included into the work.

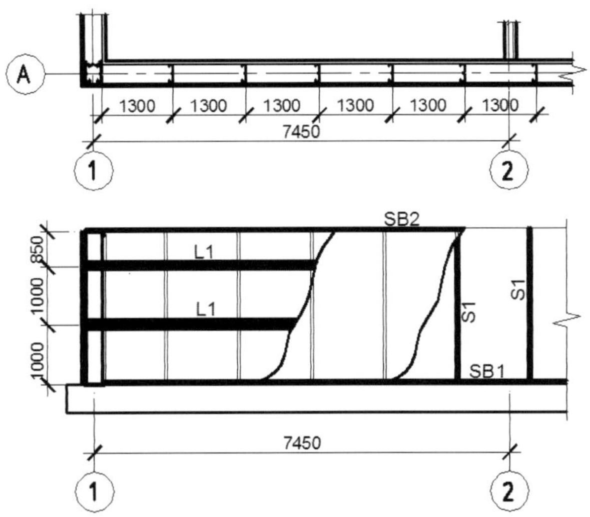

Figure 5. Layout of the thin-walled galvanized profiles of the improved frame of the building in plan and in section

In the analysis of two variants of the constructive decision of the administrative building, it was decided to use the second option. Area – 320 m^2. Technical and economic indicators of the project: cost of construction – 116,800 \$, duration – 4 months, the cost of 1 m^2 is 365 \$.

Technical and economic analysis has shown that the timing of construction in the 2nd version increased by 2 weeks, and the difference in cost is not more than 5%. The advantages of this option are: elimination of the walls instability, an increase in the racks step up to 1,300 mm, which reduces the frame material consumption. Disadvantages included an insignificant increase in the term and cost of construction due to the use of monolithic foam concrete.

4. Conclusion

In conclusion it should be noted, that the authors on the basis of personal experience, faults analysis and opinions of actual customers have improved the technology of light steel thin-walled structures in low-rise construction. Comparing the results of construction of residential houses and the administrative building, one can conclude that both options can be used in practice depending on the requirements of the customer .Application of monolithic foam concrete in frame housing construction has a positive effect on the results without a significant increase in project cost and construction time.

The application of new technologies is treated with suspicion by builders and consumers. The opinion of consumers and customers plays a huge role in the development and improvement of innovative technologies not only in low-rise construction but in other industries as well.

5. References

[1] Alkhimenko A. I., Vatin N.I., Rybakov V. A. Technology of light steel thin-walled structures. SPb: Publishing house of Polytechnical University, 2008. 27 p.

[2] Construction of buildings and structures. Light steel thin-walled structures: manual / Vatin N.I., Zhmarin E. N., Kurazhova V. G., Usanova, K. Y. SPb: Publishing house of Polytechnical University, 2012. 266 p.

[3] Zhmarin E. N., Rybakov V. A. LSTWS – a tool for the implementation of the program "Affordable and comfortable housing"//".Stroiprofil journal. 2007. № 6(60), № 7(61). P. 166-167.

[4] Nikolchak M. V., Bezhenar I. S., Rybakov V. A. Substantiation of expediency of application of technology of light steel thin-walled structures in cottage construction // XXXVIII science week SPbGPU: Proceedings of the international scientific-practical conference. 2009. Part 1. P. 219

[5] Rybakov V. A. Fundamentals of structural mechanics of light steel thin-walled structures: manual. SPb.: Izd-vo SPbGPU, 2011. 207p.

[6] Airumyan E. L., Belyi G. I. Study of the steel trusses from cold-formed profiles based on their local and general stability // Industrial and civil construction. 2010. No. 5. P. 41-44.

[7] Nazmeeva T. V. Ensuring spatial stiffness of the coating in buildings of LSTWS // Magazine of civil engineering. 2009. No. 6(8). P. 12-15.

[8] https://www.forumhouse.ru/threads/59512/ - Living in the house from LSTWS.

[9] Information on http://forum.dwg.ru.

[10] https://wiki2.org/ru/ LSTWS.

IOP Conf. Series: Journal of Physics: Conf. Series **1016** (2018) 012005 doi:10.1088/1742-6596/1016/1/012005

Control of Sit to Stand Mechanism of Assistive Device for Paraplegics

A Abdullah and Z Kausar*

Department of Mechatronics Engineering, Air University, E-9, Islamabad, Pakistan

*Email: zkau001@aucklanduni.ac.nz

Abstract: This paper focuses on position control of sit to stand mechanism of a device developed at Air University for paraplegic patients. A dynamics model of the mechanism is presented. A nonlinear robust sliding mode controller is designed to position the end effector of the mechanism correctly. Stability of switching surface is ensured and control law is devised. The position control is successfully verified through simulation results.

1. Introduction

Paraplegia refers to the paralysis of the lower half of the human body. It is the impairment affecting either motor or sensory function of lower extremities. Rehabilitation is a process that aims to restore independence and self-determination in disabled individuals. Rehabilitation process involves various kinds of therapies and assistive mechanisms that are used for training of and exercises by disabled persons.

Extensive studies have been conducted on assistive mechanisms of rehabilitation robots. These are aimed to assist the disabled person and to aid in restoring mobility of affected limbs. Such robots facilitate the patient during the training sessions[1, 2]. Various types of robotic systems designed for lower-limb rehabilitation include treadmill gait trainers[3], foot-plate-based gait trainers[4], over-ground gait trainers[5], stationary gait and ankle trainers [6] and active foot orthotics[7].

A great concern for such rehabilitation devices is the presence of a reliable control system onboard. Some previous studies insight the recent developments, challenges observed and the control strategies designed for the lower extremity exoskeletons [8].The Assistive Device for Paraplegics (ADP) under study in this work is was developed at Department of Mechatronics, Air University[9]. In this paper, we propose a control scheme for Sit to Stand (STS) mechanism of ADP. STS mechanism is responsible for assisting the patient in changing posture from sitting position to standing position and vice versa. A device for rehabilitation of the paraplegic patients has been discussed in [10]. The device aims to restore movement of the patient by applying enough amount of electrical stimulus based upon the angle of the knee joint. To improve the performance of the device, PID parameters have been optimized which improve the performance of the device. The study presented in [11, 12] focus on the development of PID controller to improve performance of the rehabilitation devices discussed in these studies.

The developed assistive device in [9]has non-linearity in dynamics behavior due to structural complexity or coupling in the system. For this purpose, an advanced control technique is proposed for STS mechanism of the device in this study which is sliding mode control (SMC). Study of sliding mode control and its variants for robotic gait rehabilitation is discussed in [13],[14]and [15]. The position control of PUMA 560 is improved using sliding mode controller and yields better positioning for the manipulator. Some of many variations of sliding mode control are given in the studies [16-19].The contribution of this paper is to design SMS control scheme for the STS mechanism and verify the applicability of the control scheme through simulation results of the system after implementation of the control system in MATLAB® Simulink.

2. Dynamics Model of the STS Mechanism

Dynamics modeling is the first step towards the control implementation. It involves mathematical modeling of the STS mechanism to describe its motion between sitting and standing position. The STS mechanism is considered as RPR (Revolute – Prismatic - Revolute) mechanism having a revolute joint followed by a prismatic and then again a revolute joint. A free body diagram of the STS mechanism is shown in Figure 1 (b) whereas a CAD model is shown in figure 1(a).

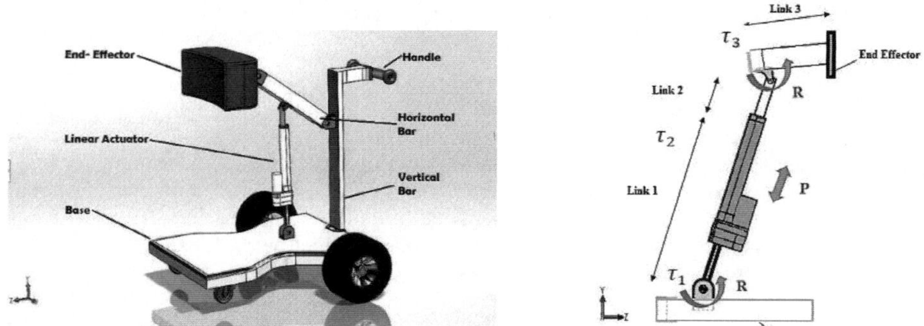

(a)

(b)

Figure 1. (a) The CAD model of Assistive Device for Paraplegics, (b) Free body Diagram of STS mechanism[9]

The equations of motion are derived using Lagrange approach. The Lagrange method refers to the total energy of the system. The equations of motion of the STS system derived are given in (1) to (3).

Link-1:
$$\tau_1 = \left(m_1 l_1^2 + Izz_1 + m_2 d_2^2 + Iyy_2 + m_3 l_{C_3}{}^2 + Ixx_3 S_3{}^2 + Iyy_3 C_3{}^2\right)\ddot{\theta}_1 + 2m_2 d_2 \dot{d}_2 \dot{\theta}_1 - m_2 d_2 \dot{\theta}_1{}^2 + (m_1 l_1 + m_2 d_2) g C_1 + m_3 g l_{C_3}(S_1 S_3 - C_1 C_3) \quad (1)$$

Link-2:
$$\tau_2 = (m_2 + m_3)\ddot{d}_2 - m_2 d_2 \dot{\theta}_1{}^2 + m_2 g S_1 \quad (2)$$

Link-3:
$$\tau_3 = \left(m_3 l_{C_3}{}^2 + Izz_3\right)\ddot{\theta}_3 - (Ixx_3 S_3 C_3 - Iyy_3 C_3 S_3)\dot{\theta}_1{}^2 + m_3 g l_{C_3}(C_1 C_3 - S_1 S_3) \quad (3)$$

Where τ_i is the torque produced at ith joint,,d_2 is displacement of linear actuator,C_i,$S_i$is $Cos\ (\theta_i)$,$Sin\ (\theta_i)$, m_i is mass of the ith link, I moment of inertia of the ith link along respective axis and l is the length of ith link. Link-1 is actuated by first revolute joint, link-2 is actuated by the prismatic joint and link-3 by the second revolute joint.The parameters like mass and moment of inertia are computed from the 3D CAD model by assigning a material to the model. The link lengths are taken from the actual system.

3. Sliding Mode Control (SMC)

IOP Conf. Series: Journal of Physics: Conf. Series **1016** (2018) 012005 doi:10.1088/1742-6596/1016/1/012005

SMC belongs to the family of Variable Structure Controllers (VSC). Design of VSCs involve two major phases. In the first phase, a switching function is designed which is generally called a sliding surface. The second phase is the design of the control law.

3.1. Sliding Surface design

Ideally, once intercepted, the switched control maintains the plant's state trajectory on the surface to slide along until it reaches the desired point. This is also known as the sliding phase. This surface acts as guide for plant's states to fall on to it and slide along until they reach the desired point. If x = current position, and x_d = desired position then the error $\tilde{x}$can be given as (4):

$$\tilde{x} = x - x_d \qquad (4)$$

The sliding surface $s(x,t)$can be given as (5):

$$s(x,t) = (\frac{d}{dt} + \lambda)^{(n-1)}\tilde{x} = 0 \qquad (5)$$

Where n is the order of the system and λ is a constant positive gain. For n=2, the sliding surface proposed becomes as given by (6):

$$s(x,t) = \dot{\tilde{x}} + \lambda\,\tilde{x} = 0 \qquad (6)$$

3.2. Control Law

The sliding surface ensures that once the system's states are on the surface, they will tend to take the system to the desired point. But sliding surface only provides the switching information. The phase in which the state start moving towards the surface is known as the reaching phase. In order to ensure that the state reaches the surface, the sliding surface must fulfill the reach-ability condition. Reach-ability condition states that

$$s\dot{s} < 0 \qquad (7)$$

General form of the control law is stated as (8) where U is the control input of the system and $U_0$is initial input of the system:

$$U = -U_0\, sgn(s) \qquad (8)$$

Now for the reachability condition we calculated derivative of the sliding surface as given in (9):

$$\dot{s} = \ddot{x} - \ddot{x}_d + \lambda\dot{\tilde{x}} \qquad (9)$$

The STS mechanism is actuated by a linear actuator at link 2 shown in figure 1(b). The torque input $\tau_2 = U$ for link 2 . We rearranged (2) and resulting equation is (10).

$$(m_2 + m_3)\ddot{d}_2 = U + m_2 d_2 \dot{\theta}_1^{\,2} - m_2 g \sin\theta_1 \qquad (10)$$

After rearrangement of equation (10) we find an equivalent control input $\widehat{U}$ as given in (11).

$$\widehat{U} = (m_2 + m_3)\ddot{x}_d - \lambda\dot{\tilde{x}}(m_2 + m_3) - m_2 d_2 \dot{\theta}_1^{\,2} + m_2 g \sin\theta_1 \qquad (11)$$

The overall control law can now be written as (12):

$$U = \widehat{U} - k\,\mathrm{sgn}(s) \qquad (12)$$

For the reach-ability condition we deduced the product $s\dot{s}$ which results in equation (13):

$$\dot{s}.s = \left(\frac{1}{m_2 + m_3}[-k\,\mathrm{sgn}(s)]\right)s \qquad (13)$$

If we choose k such that the overall sign of the product remains negative then the condition for reach-ability is satisfied.

4. Simulation results

Simulink Model for the dynamics model and the controller formulated was developed. The gains of SMC were tuned after the successful implementation of sliding surface and control law on the plant. The tuning of gains is crucial for acquiring good results. For the simulation purpose the desired position for the STS mechanism was taken to be 0.2 m from the start point which is maximum length produced by actual linear actuator. We assumed that the STS mechanism was initially at sitting position. We desired the STS mechanism to move the patient from sitting position to the standing position. A young healthy person, to stand up from a sitting position, takes about 2–2.5 seconds. So the desired speed was calculated accordingly. The simulation results of SMC for STS mechanism are shown in figures 2-5. Figure 2&3 shows motion of two revolute joints while figure 4 verifies the tracking of the desired trajectory of the link-2 end-effector. The control input shown in figure 5 is such that it can be implemented in a real system.

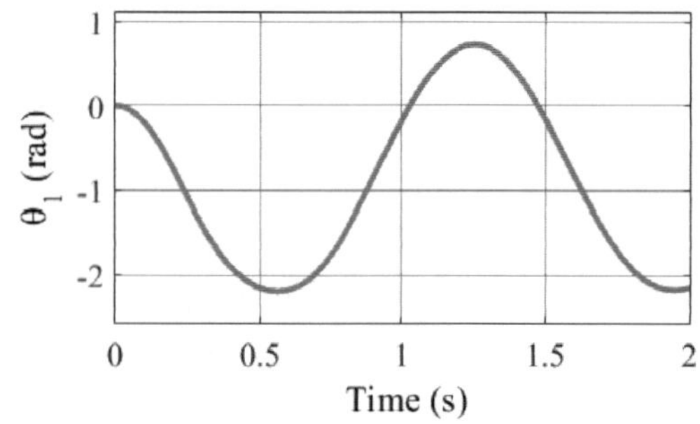

Figure 2. Plot for joint angle θ_1

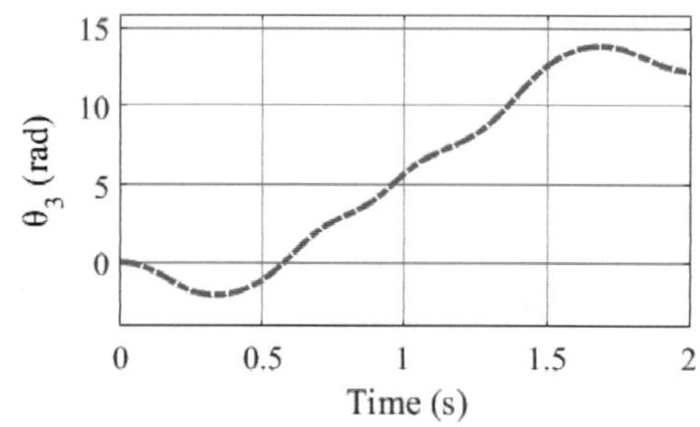

Figure 3. Plot for joint angle θ_3

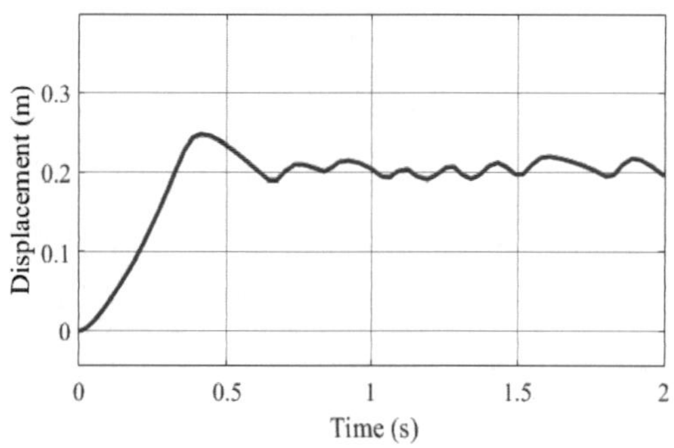

Figure 4. Plot for displacement ‘d2’ for link 2

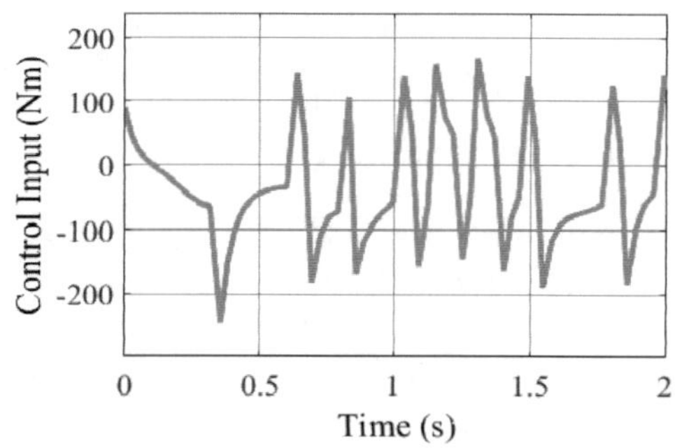

Figure 5. Control Input Plot

5. Conclusions

In this study, Sliding Mode Controller was proposed for the control of the position of the STS mechanism of the assistive device for paraplegic patients. The dynamics modeling of the STS mechanism show that the system is highly non linear. Thus the simulations were performed using SMC as the non linear controller to control the system. It is shown in simulation results that SMC is able to control the device. The non linear controller performed well in switching the requirements and controlling the motion of the STS mechanism. An oscillation in the position is caused which may be removed and the results may be compared with PID controller as a future work.

References

[1] I. Díaz, J. J. Gil, and E. Sánchez, "Lower-limb robotic rehabilitation: literature review and challenges," *Journal of Robotics,* vol. 2011, 2011.

[2] O. Salah, A. Ramadan, S. Sessa, and A. A. Abo-Ismail, "A systematic approach for design a low-cost mobility assistive device for elderly people," *Int J Med Health, Pharm Biomed Eng,* vol. 5, pp. 36-41, 2011.

[3] M. Pietrusinski, I. Cajigas, Y. Mizikacioglu, M. Goldsmith, P. Bonato, and C. Mavroidis, "Gait rehabilitation therapy using robot generated force fields applied at the pelvis," in *Haptics Symposium, 2010 IEEE*, 2010, pp. 401-407.

[4] J. Yoon, B. Novandy, C.-H. Yoon, and K.-J. Park, "A 6-DOF gait rehabilitation robot with upper and lower limb connections that allows walking velocity updates on various terrains," *IEEE/ASME Transactions on Mechatronics,* vol. 15, pp. 201-215, 2010.

[5] K.-H. Seo and J.-J. Lee, "The development of two mobile gait rehabilitation systems," *IEEE transactions on neural systems and rehabilitation engineering,* vol. 17, pp. 156-166, 2009.

[6] C. Schmitt and P. Métrailler, "The Motion Maker™: a rehabilitation system combining an orthosis with closed-loop electrical muscle stimulation," in *8th Vienna International Workshop on Functional Electrical Stimulation*, 2004, pp. 117-120.

[7] G. S. Sawicki and D. P. Ferris, "A pneumatically powered knee-ankle-foot orthosis (KAFO) with myoelectric activation and inhibition," *Journal of neuroengineering and rehabilitation,* vol. 6, p. 23, 2009.

[8] B. Chen, H. Ma, L.-Y. Qin, F. Gao, K.-M. Chan, S.-W. Law, *et al.*, "Recent developments and challenges of lower extremity exoskeletons," *Journal of Orthopaedic Translation,* vol. 5, pp. 26-37, 2016.

[9] M. Khan and Z. Kausar, "Kinematics Modeling and Mechanism Configuration Selection of a Mechatronic Assistive Device for Paraplegic Patients," in *Proceedings of the 5th International Conference on Mechatronics and Control Engineering*, 2016, pp. 27-32.

[10] F. Previdi, M. Ferrarin, I. Carpinella, and S. Savaresi, "Modelling and control of a device for rehabilitation of paraplegic patients," in *American Control Conference, 2007. ACC'07*, 2007, pp. 2030-2035.

[11] L. Khan, M. Irfan, and H. Ali, "Implementation of GA in designing control system for paraplegic walking movement using walker," *World Applied Sciences Journal,* vol. 7, pp. 1164-1175, 2009.

[12] K. Suzuki, G. Mito, H. Kawamoto, Y. Hasegawa, and Y. Sankai, "Intention-based walking support for paraplegia patients with Robot Suit HAL," in *Climbing and Walking Robots*, ed: InTech, 2010.

[13] S. Hussain, "Design and Assist-as-Needed Control of an Intrinsically Compliant Robotic Orthosis for Gait Rehabilitation," Masters, Department of Mechanical Engineering, The University of Auckland, Auckland Newzeland, 2012.

[14] S. Alireza Moezi, M. Rafeeyan, and S. Ebrahimi, "Sliding mode control of 3-RPR parallel robot on the optimal path using cuckoo optimization algorithm," *Modares Mechanical Engineering,* vol. 15, 2015.

[15] J. M. ZARYN and J. E. KUREK, "Comparison of sliding mode control and decoupled sliding mode control of robot Puma 560," 2003.

[16] M. L. Corradini, V. Fossi, A. Giantomassi, G. Ippoliti, S. Longhi, and G. Orlando, "Discrete time sliding mode control of robotic manipulators: Development and experimental validation," *Control Engineering Practice,* vol. 20, pp. 816-822, 2012.

[17] S. Islam and X. P. Liu, "Robust sliding mode control for robot manipulators," *IEEE Transactions on Industrial Electronics,* vol. 58, pp. 2444-2453, 2011.

[18] T. Sun, H. Pei, Y. Pan, H. Zhou, and C. Zhang, "Neural network-based sliding mode adaptive control for robot manipulators," *Neurocomputing,* vol. 74, pp. 2377-2384, 2011.

[19] M. Zeinali and L. Notash, "Adaptive sliding mode control with uncertainty estimator for robot manipulators," *Mechanism and Machine Theory,* vol. 45, pp. 80-90, 2010.

IOP Conf. Series: Journal of Physics: Conf. Series **1016** (2018) 012006 doi:10.1088/1742-6596/1016/1/012006

A spatial registration method for navigation system combining O-arm with spinal surgery robot

H Bai[1,2], G L Song[1] Y W Zhao[1] X Z Liu[1,2] and Y X Jiang[1]

[1] Shenyang Institute of Automation, Chinese Academy of Sciences, China.
[2] Northeastern University, China.

Email: songgl@sia.cn

Abstract. The minimally invasive surgery in spinal surgery has become increasingly popular in recent years as it reduces the chances of complications during post-operation. However, the procedure of spinal surgery is complicated and the surgical vision of minimally invasive surgery is limited. In order to increase the quality of percutaneous pedicle screw placement, the O-arm that is a mobile intraoperative imaging system is used to assist surgery. The robot navigation system combined with O-arm is also increasing, with the extensive use of O-arm. One of the major problems in the surgical navigation system is to associate the patient space with the intra-operation image space. This study proposes a spatial registration method of spinal surgical robot navigation system, which uses the O-arm to scan a calibration phantom with metal calibration spheres. First, the metal artifacts were reduced in the CT slices and then the circles in the images based on the moments invariant could be identified. Further, the position of the calibration sphere in the image space was obtained. Moreover, the registration matrix is obtained based on the ICP algorithm. Finally, the position error is calculated to verify the feasibility and accuracy of the registration method.

1. Introduction

In the past two decades, minimally invasive spine surgery (MISS) has been increasingly applied and drawn much attention in the treatment of spinal disorders[1]. The development of image technology has led to a more significant development of minimally invasive spinal surgery. Not only can MISS minimize injury to paraspinal back muscles, connective tissues, and joints but it can also decrease the amount of bleeding, infection, hospital stay, and postoperative pain[2].It decreases the incidence of complications and approach-related morbidity and mortality associated with conventional open surgery. The traditional open spine surgery has gradually been replaced with MISS. According to the reports[1], the number of MISS conducted in 2010 accounted for 1/6 of the total number of all spine surgeries in the United States and 1/3 in 2016, which is anticipated to be more than 1/2 in 2020.

Although MISS has the above advantages, it still faces to the following problems: (1) Surgeon's surgical vision is limited. (2) With the increasing number of surgery, the surgeon's intra-operative radiation exposure will accumulate. (3) The lack of surgical instrument's real-time position, reduces the safety of surgical procedures.

In order to solve the above problems, some scholars have proposed that the robot navigation technology could assist surgeons. A study assessed the feasibility and clinical value of robot assisted navigation drilling for pedicle screw placement and the results confirmed that robot assisted surgery could increase the quality of percutaneous pedicle screw placement[3]. The accuracy of guidance systems in screw insertion procedures is particularly notable in anatomically difficult cases, such as scoliosis correction surgeries, with a reported 6-fold reduction in perforation rates and mean insertion angle errors compared with conventional methods[4]. In the meta-comparison of navigation-assisted versus conventional screw placement covering 28 clinical, 14 cadaveric, and 1 model-based study, Tian [5], reported a higher overall accuracy of screw position in navigation-assisted procedures.

The Rosa spine robot[6], which has been used clinically, combines robots and navigation systems to obtain the pose of surgical instruments and patients through two IR cameras. Under the assistance of intra-operative images, the surgical robot is guided to the surgical position and finally assists surgeons place pedicle screws. The navigation system of Rosa spine robot is through the optical imaging system to achieve spatial positioning and such systems are vulnerable to the effect of intra-operative objects' shelter, the surrounding light and metallic objects mirror reflection. What' s more, the Renaissance guidance system[7], uses operating forceps or Hover-T to fix a miniature parallel robot on the patient's spine. The C-arm takes anterior–posterior and lateral intra-operative fluoroscopic images of the patient and transfers them to the workstation system for automatic image registration. But the specially designed jig with three marker holes need to be fixed to the patient's body, causing additional trauma to the patient.

In order to solve the above problems, this paper proposes a robotic navigation system based on O-arm. One of the major problems of this system is to registration of patient space and intra-operation image space. This study mainly introduces the registration method of the system.

2. Material and method

2.1. Material

This study designed a calibration phantom to assist spatial registration and error calculations. Taking the registration into account, the main criterion given for the phantom was the need to provide multiple surfaces for placement the calibration sphere[8]. The solution for this criterion was to design a 70mm × 70mm × 80mm cube calibration phantom. The four faces of which are accurately placed with calibration holes, each with 16 calibration holes on each plane to provide sufficient test volume. Figure 1 illustrates the used calibration phantom. The calibration holes were machined 15mm apart, and the position error for the distance horizontally and vertically was within ± 0.02 mm. The diameter and depth of the calibration hole is 2.5 mm, which ensures that the calibration sphere is exactly filled in the hole.

The use of O-arm has been increasing in recent years, because the medical staff can leave the operating theatre to decrease their exposure to radiation, during an O-arm 3D image acquisition[9]. The O-arm is a mobile intra-operative computed tomography imaging system optimized for bony structures in spinal and orthopedic surgery [10]. Pixel size is 0.424 × 0.424 mm within a slice thickness of 0.625 mm. Figure 2 illustrates the calibration phantom and the spinal phantom with the O-arm.

Figure 1. The present calibration phantom.

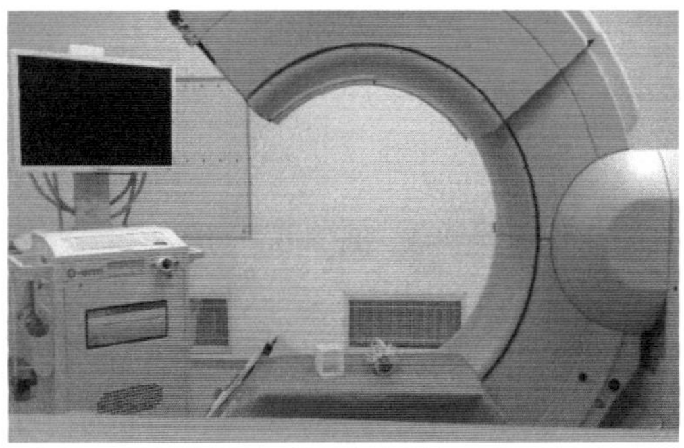

Figure 2. The two phantom with O-arm.

2.2. Methods

Connecting the calibration phantom and the Stäubli robot through the connecting rod. Figure 3 illustrates the way of connection between the calibration phantom and robot. The position of the calibration sphere in the patient space is obtained by the kinematics of the robot. Scanning the calibration phantom and the spinal phantom together by the O-arm to obtain CT slices containing both phantoms. The robot and the connecting rod are kept at a certain distance from the scanning area to prevent serious metal artifacts in the CT slices, during the O-arm scanning.

The material of the calibration sphere placed in the phantom is metal. The gray-scale value of the calibration sphere is very different from the gray-scale value of the calibration phantom in the CT slices. Therefore, it is easy to identify the calibration sphere in the slices. Although the O-arm from the multi-angle fluorescence projection to the CT slice is very fast, there are metal artifacts around calibration spheres in the CT slices, due to beam hardening. Figure 4 illustrates one of the CT slices containing metal artifacts. Metal artifacts result in dimensional distortion of the slices of the calibration sphere, which require to reduce to improve the accuracy of image feature extraction.

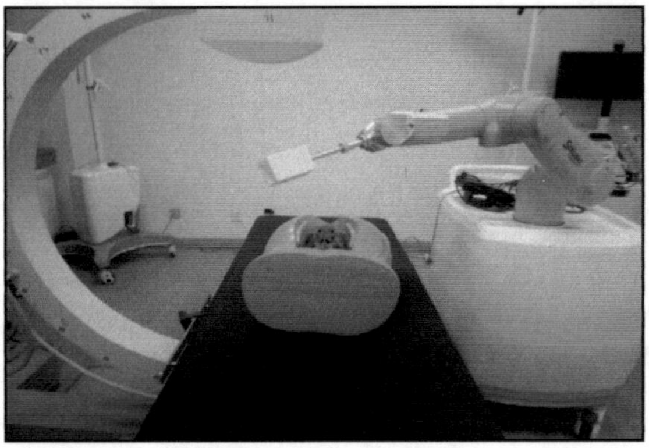

Figure 3. Connection between the phantom and robot.

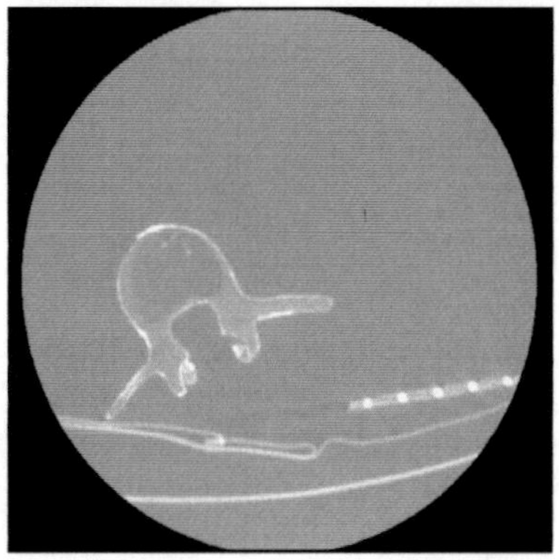

Figure 4. CT slice with metal artifacts.

Metal artifact reduction.

Most metal artifact reduction (MAR) researches focused on the development of methods which use raw CT scan data. But in most applications, the raw data is not available and researchers can only infer the true results from the containing artifacts filtered back projection (FBP) images. In order to solve this problem, Naranjo[11], made the FBP image be converted to the polar coordinate system, before the image processing. And the streaking artifacts are converted into vertical rays, which can be easily removable using a morphological filter. This method was used for metal artifact reduction because of the radial artifacts that occur around the calibration sphere in the CT slices, resulting in its shape being not a standard circle.

The purpose is to preserve the original structure of the initial image as much as possible while reducing artifacts. The result of closing–opening average filter, as in equation (1) is the average of the results of the closing and opening operators. The maximum gray-scale value changes the smallest and the noise gray-scale value changes the maximally. That is the best reduction results.

$$\mathrm{COA}_{\lambda} = \frac{\gamma_{B_{\lambda}}(f) + \varphi_{B_{\lambda}}(f)}{2} \tag{1}$$

The results of the morphological filters are associated not only with the type of filter but also with the correct selection of the shape and size of structuring element (SE). To overcome this problem a distance adaptive size of the SE is used. Let $\lambda = 2 \square 20$. Using the center point of the artifact as the origin of the coordinate transformation. The transformation into polar coordinates is defined by the following equation (2):

IOP Conf. Series: Journal of Physics: Conf. Series **1016** (2018) 012006 doi:10.1088/1742-6596/1016/1/012006

$$\rho=\sqrt{(x-x_c)^2+(y-y_c)^2}\ ,\qquad 0\le\rho\le\rho_{\max}$$
$$\theta=\arctan\left(\frac{y-y_c}{x-x_c}\right),\qquad 0\le\theta\le 2\pi \tag{2}$$

Defining (x_c, y_c) as the origin of the streaking rays, that is, taking the streaking source as the focus of the transformation.

Using line detection and Hough transform to extract striped artifacts[12]. Figure 5 is a flow chart for obtaining the center coordinates of the metal object.

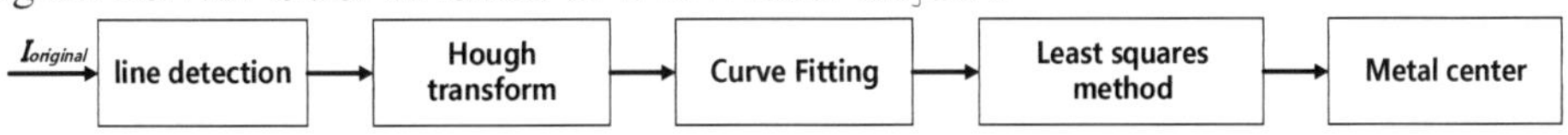

Figure5. The flow chart for obtaining the metal center coordinates.

The basic steps are as follows:

(1) Extracting stripe artifacts by line detection and Hough transform to find the center coordinates of metal objects.

(2) Converting the initial image from the Cartesian coordinate system to the polar coordinate system.

(3) In the polar coordinate system using closing–opening filter to remove artifacts.

(4) The filtered image is converted from the polar coordinate system to the Cartesian coordinate system to obtain $\mathbf{I}_{filtered}$.

(5) The initial image is processed by threshold to get I_{msk}.

(6) The final image is obtained by combining the initial image with the filtered image, as in equation (3).

$$I_{output}=I_{original}\times I_{filtered}\times\left(1-I_{msk}\right) \tag{3}$$

Circle recognition. In order to improve the speed of image feature recognition and shorten the operation time, the moment invariants of the images were used to represent the shape features of the sectional circles of the metal spheres in the CT slices. Moment invariants are very useful tools for pattern recognition. They were firstly derived by Hu in 1963 which were frequently used as features for shape recognition [13] and Hu's invariants are usable based on region. Then Chen [14] has proposed improved moment invariants based on boundary in 1993. Both the above-mentioned features are not valuable based on both regions and boundaries simultaneously. So we use the united moment invariants derived by Sun[15] which can be applied based on both region and boundary. Sun derived the following moment invariants, as in equation (4):

$$\begin{cases}\theta_1=\sqrt{\phi_2}\,/\,\phi_1 & \theta_2=\phi_6\,/\,\phi_1\,\phi_4\\ \theta_3=\sqrt{\phi_5}\,/\,\phi_4 & \theta_4=\phi_5\,/\,\phi_3\,\phi_4\\ \theta_5=\phi_1\,\phi_6\,/\,\phi_2\,\phi_3 & \theta_6=\left(\phi_1+\sqrt{\phi_2}\right)\phi_3\,/\,\phi_6\\ \theta_7=\phi_1\,\phi_5\,/\,\phi_3\,\phi_6 & \theta_8=\left(\phi_3+\phi_4\right)/\sqrt{\phi_5}\end{cases} \tag{4}$$

Where ϕ_i are Hu' s moment invariants.

Sphere center calculation. Through identifying the position of the circle in the CT slices by the above eight formulas, but the problem is lacking the depth information about the center of the sphere. We need to obtain the radius of the circle and the radius of the sphere to compute the depth information. It is known that any section of the sphere must be circular, and the circle of the center of

IOP Conf. Series: Journal of Physics: Conf. Series **1016** (2018) 012006 doi:10.1088/1742-6596/1016/1/012006

the circle and the core of the connection will be perpendicular to the cross section. Through the experiment, we found that the position of the center of the same sphere was almost constant. So it was possible to use the Pythagorean theorem to find the distance from the cross section to the center of the sphere. Because the direction of the center of the sphere is uncertain, two distances were obtained with depth information for each section. The trend of the radius changing can exclude the false distance and keep the true distance. The center of the sphere is represented by the average of the depth information about a sphere in the intra-operation image space.

Spatial registration algorithm. ICP algorithm has the advantages of fast convergence speed. The spatial transformation parameter of point set is the spatial mapping parameter of image. The registration algorithm consists of two steps: preregistration and precise registration.

A good initial value is important to the iterative algorithm, because it can prevent the optimal solution from falling into the local minimum solution and reduce the time consumption of the registration algorithm. The initial registration is used to provide an iteration initial value near the global optimal solution. Finding the pre-selected four non-coplanar mark points, through these four points to complete the pre-registration. The four marked points in patient space and intra-operation image space form two matrices P_R^M and P_i^M , respectively, as shown in equation (5) and (6).

$$P_R^M = \begin{bmatrix} x_R^1 & x_R^2 & x_R^3 & x_R^4 \\ y_R^1 & y_R^2 & y_R^3 & y_R^4 \\ z_R^1 & z_R^2 & z_R^3 & z_R^4 \\ 1 & 1 & 1 & 1 \end{bmatrix} \tag{5}$$

$$P_i^M = \begin{bmatrix} x_i^1 & x_i^2 & x_i^3 & x_i^4 \\ y_i^1 & y_i^2 & y_i^3 & y_i^4 \\ z_i^1 & z_i^2 & z_i^3 & z_i^4 \\ 1 & 1 & 1 & 1 \end{bmatrix} \tag{6}$$

The pre-registration matrix ${}^iT_r^1$ is shown in equation (7).

$${}^iT_r^1 = P_i^M (P_r^M)^{-1} \tag{7}$$

After the pre-registration matrix is obtained, the point cloud in patient space P_r can be converted to $P_r^{'}$ by pre-registration matrix ${}^iT_r^1$, as shown in equation (8). Then the pre-registration is completed.

$$P_r^{'} = {}^iT_r^1 P_r \tag{8}$$

After pre-registration, the relationship between the patient space and the intra-operation image space can be calculated. ICP registration algorithm is used to accomplish precise registration, which was first proposed by Besl and Mckay[16], the essence of which is that based on least squares registration. The rotation matrix R and the position vector P are updated until equation (9) is satisfied. Finally, the registration matrix ${}^iT_r^2$ was obtained.

$$f(R,T) = \sum_{k=1}^{n} \left\| P_i^{k'} - (RP_r^k + T) \right\|^2 = \min \tag{9}$$

3. Results

After obtaining the CT slices, the DICOM format is converted to BMP format, which will increase the speed of the next processing. The BMP format image is Gaussian filter, Canny edge detection and contour extraction, calculate the moment invariants of each contour. When the condition is met, the contour ellipse is fitted again, and Figure 6 shows the result of the recognition based on the moment invariants, where the identified ellipse is highlighted in red. The elliptical maximum inscribed circle is defined as the sectional circle of the calibration sphere, so the circle radius is elliptical short axis, and the center of the circle is the center of the ellipse.

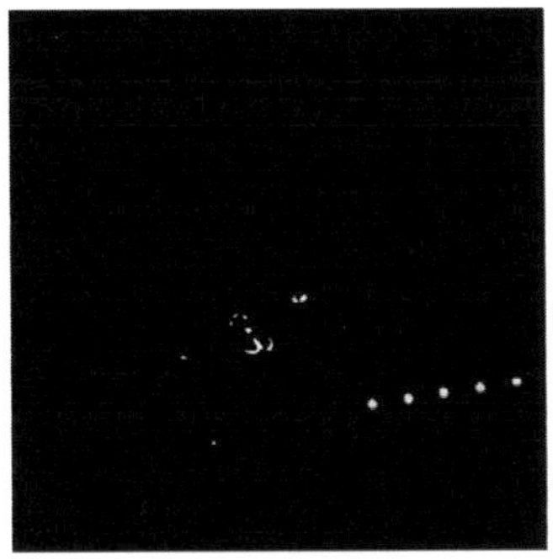

Figure 6. The red oval for the fitting results.

By moving the robot, scanning again and the second set of data was obtained to calculate the error. The center position （X_r^e ,Y_r^e ,Z_r^e） in the patient space is multiplied by the registration matrix ${}^{i}T_r^2$. The calculated results are compared with the centroid position （X_i^e ,Y_i^e ,Z_i^e） in the image space. The error analysis is based on the position error between the calculated sphere center coordinates and the sphere center coordinates obtained by the robot kinematics, as shown in equation (10). Figure 7 illustrates the histogram of the position error.

$$Error = \left|(calculated\ value) - (true\ value)\right| \tag{10}$$

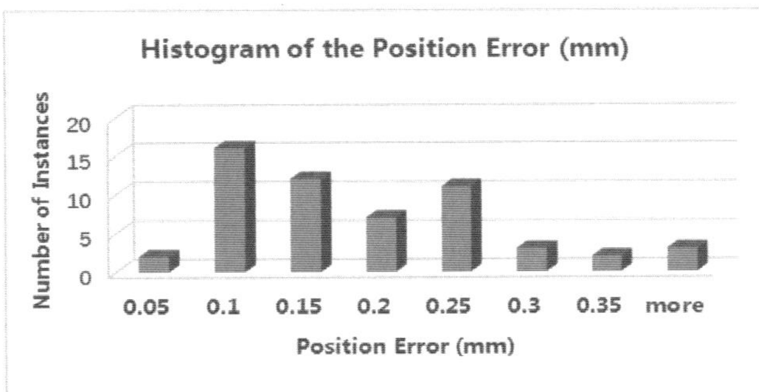

Figure 7. The histogram of the position error.

4. Conclusions

In this study, metal artifact reduction, feature recognition based on moment invariants and spatial registration method based on ICP were used to calculate the position of the same group of spheres in different spaces, and the registration of intra-operation image space and patient space was realized. In order to solve the intra-operative combination of the O-arm and the robot navigation problem, without increasing the patient's additional trauma, the idea of mechanical positioning navigation is proposed. A new way of consideration is provided for the surgery robot navigation technology.

But there are still the following questions:

First, the experimental process need to maintain the relative fixation of the vertebral body, spinal movement or other deformation will lead to a decline in navigation accuracy.

Second, because the validity of the algorithm is affected by the reconstructed FBP image quality and the effect of the morphological filter, the image effect of the artifacts that are not serious is not significant, and the adaptability of the morphological filter will be strengthened.

It is necessary to research the above issues in later studies.

References

[1] Huang T J, Kim K T, Nakamura H, Yeung A T and Zeng J 2017 *Biomed Research International* **2017**(5) pp 1-4
[2] Kim J S, Härtl R and Mayer H M 2016 *Biomed Research International* **2016** pp 5048659

[3] Hongwei W, Yue Z, Jun L, Jianda H and Liangbi X 2015 *Indian Journal of Orthopaedics* **49**(4) pp 452-457
[4] Keric N, Doenitz C, Haj A, Rachwalczyzewicz I, Renovanz M, Dma W, et al. 2017 *Neurosurgical Focus* **42**(5) pp E11
[5] Tian N F, Huang Q S, Zhou P, Zhou Y, Wu R K, Lou Y, et al. 2011 *European Spine Journal* **20**(6) pp 846-859
[6] Lefranc M and Peltier J 2015 *Journal of Robotic Surgery* **9**(4) pp 331-338
[7] Sukovich W, Brinkdanan S and Hardenbrook M 2006 *International Journal of Medical Robotics & Computer Assisted Surgery* **2**(2) pp 114-22
[8] Koivukangas T and Katisko J P A 2010 *5th Cairo International Biomedical Engineering Conference* (Cairo, Egypt) pp 57-60
[9] Pitteloud N, Gamulin A, Barea C, Damet J, Racloz G and Sans-Merce M 2016 *European Spine Journal* **26(3)** pp 1-7
[10] Koivukangas T, Katisko J P and Koivukangas J P 2011 *Conf Proc IEEE Eng Med Biol Soc* **2011**(4) pp 2148-2151
[11] Naranjo V, Lloréns R, Alcañiz M and López-Mir F 2011 *Computer Methods & Programs in Biomedicine* **102**(1) pp 64-74
[12] Naranjo V, Llorens R, Paniagua P and Albalat S 2009 *International Work-Conference on the Interplay Between Natural and Artificial Computation* (Springer, Berlin, Heidelberg) pp 11-19
[13] Hu M K 1962 *IRE transactions on information theory* **8**(2) pp 179-187
[14] Chen C C 1993 *Pattern Recognition* **26**(5) pp 683-686
[15] Sun Y, Liu W and Wang Y 2003 *IEEE International Conference on Robotics, Intelligent Systems and Signal Processing, 2003. Proceedings* (Changsha, Hunan, China) 1 pp 88-93
[16] Besl P J and McKay N D 1992 *IEEE Transactions on pattern analysis and machine intelligence* **14**(2) pp 239-256

Acknowledgments

This work was supported in part by National Natural Science Foundation of China (Grant No. 61333019 and 61703394).

Modelling and Closed-Loop System Identification of a Quadrotor-Based Aerial Manipulator

Chioniso Dube and Jimoh O Pedro

School of Mechanical, Industrial and Aeronautical Engineering, University of the Witwatersrand, Johannesburg, South Africa

E-mail: chioniso.kuchwa-dube@wits.ac.za

Abstract. This paper presents the modelling and system identification of a quadrotor-based aerial manipulator. The aerial manipulator model is first derived analytically using the Newton-Euler formulation for the quadrotor and Recursive Newton-Euler formulation for the manipulator. The aerial manipulator is then simulated with the quadrotor under Proportional Derivative (PD) control, with the manipulator in motion. The simulation data is then used for system identification of the aerial manipulator. Auto Regressive with eXogenous inputs (ARX) models are obtained from the system identification for linear accelerations $\ddot{X}$ and $\ddot{Y}$ and yaw angular acceleration $\ddot{\psi}$. For linear acceleration $\ddot{Z}$, and pitch and roll angular accelerations $\ddot{\theta}$ and $\ddot{\phi}$, Auto Regressive Moving Average with eXogenous inputs (ARMAX) models are identified.

1. Introduction

Aerial manipulators have the ability not only to move in three-dimensional space, but to physically interact with their environment. Potential applications for aerial manipulators include inspection by contact of structures such as bridges, repair and maintenance of high-voltage overhead power lines, precision agriculture, and geological sampling of volcanic rock. While research has been conducted in the modelling and control of aerial manipulators; system identification of aerial manipulators has not received attention. System identification has, however, been performed for a quadrotor alone without a manipulator. For instance, Zhang et al., [1] surveyed various modelling and identification methods employed for quadrotors.

A number of authors first derived the quadrotor model using either the Newton-Euler or Euler-Lagrange formulations and then identified parameters such as thrust and drag parameters. Derafa et al., [2] identified the inertia matrix, dynamic friction coefficients and translational drag coefficients for a quadrotor. They created a rotational pendulum rig by suspending the quadrotor on wires to measure the inertia, and translational drag parameters. Dong et al., [3] used the Newton-Euler formulation to develop the dynamic model of the quadrotor. They calculated the thrust and aerodynamic drag of the quadrotor and used a three-wire pendulum set-up to measure the inertia of the quadrotor. Elsamanty et al., [4] also used the Newton-Euler formulation for the quadrotor dynamic model. They used a Computer-Aided Design (CAD) model to calculate the mass moment of inertia and geometrical parameters of a quadrotor. They then identified the propeller angular velocity using a microphone as a tachometer. Thrust and drag moment were identified using a lever arm test rig. Araar et al., [5] identified the thrust factor and drag factor of a quadrotor using a didital scale and tachometer. They obtained the inertia matrix from the CAD model. Hatamleh et al., [6] performed a simulation study to

IOP Conf. Series: Journal of Physics: Conf. Series **1016** (2018) 012007 doi:10.1088/1742-6596/1016/1/012007

investigate the dynamic model parameters of a quadrotor in the presence of noisy feedback signals. They compared three estimation methods and found that the hybrid of artificial neural network and iterative bi-section shooting had a higher accuracy in estimating the parameters.

Black-box system identification has been performed by various authors, using both simulated quadrotors and physical quadrotors. Lee et al., [7] used the Prediction Error Method (PEM), to estimate a linearised discrete-time state-space model for a quadrotor. They used a six Degree of Freedom (DoF) test-rig to perform tests in addition to actual flight tests. Stanculeanu and Borangiu, [8] used a discrete-time linear state-space model for the quadrotor. They performed black-box system identification for the quadrotor in near hover flight using the PEM. Schreurs et al., [9] developed a simulation model of a quadrotor and used the PEM for system identification from which they obtained a 12th-order Auto Regressive Moving Average with eXogenous inputs (ARMAX) model. Bergamasco et al., [10] performed flight experiments for a hovering quadrotor and logged the flight data. They used the flight data to derive a continuous-time model using subspace model identification methods. Pedro and Crouse, [11] performed system identification for a simulated quadrotor using radial basis function neural networks. Black-box system identification for a quadrotor, using dynamic neural networks was also performed by Pedro and Dangor [12].

While previous studies have focused on the system identification of quadrotors, none have attempted the system identification of quadrotor-based aerial manipulators. This paper therefore presents the system identification of a quadrotor-based aerial manipulator with a three-DoF manipulator. The remainder of this paper is structured as follows: section 2 presents the dynamic model of the quadrotor-based aerial manipulator. Section 3 describes the system identification performed and results obtained. Conclusions are then presented in section 4.

2. Aerial Manipulator Dynamic Modelling

The quadrotor model is developed using the Newton-Euler formulation. For the manipulator, the Recursive Newton-Euler formulation is used. By modelling the manipulator with the quadrotor as its floating base, the two models are combined to form the model of the aerial manipulator.

2.1. *Quadrotor Model*

Figure 1 shows the quadrotor system reference frames: the earth-inertial frame (E-frame) represented by (${}^{E}X$, ${}^{E}Y$, ${}^{E}Z$) and the quadrotor body frame (B-frame) represented by (${}^{B}x$, ${}^{B}y$, ${}^{B}z$). The inertial and body frames are related by three successive rotations about the z, y and x axes. These are given by the Euler (z y x) rotations, where roll ϕ is the rotation about the x-axis, pitch θ is the rotation about the y-axis, and yaw ψ is the rotation about the z-axis. The rotation from the body frame to the inertial frame is given by the rotation matrix ${}^{E}\boldsymbol{R}_{B}$. ${}^{E}\boldsymbol{T}_{B}$ transforms the angular velocities in the body frame to the time variation of the Euler angles. In matrix form, the Newton-Euler formulation for the quadrotor is:

$$\begin{bmatrix} {}^{B}\boldsymbol{F} \\ {}^{B}\boldsymbol{\tau} \end{bmatrix} = \begin{bmatrix} m_Q\mathbb{I}_3 & \mathbb{O}_3 \\ \mathbb{O}_3 & \boldsymbol{I}_Q \end{bmatrix} \begin{bmatrix} {}^{B}\dot{\boldsymbol{V}} \\ {}^{B}\dot{\boldsymbol{\omega}} \end{bmatrix} + \begin{bmatrix} {}^{B}\boldsymbol{\omega} \times m_Q\, {}^{B}\boldsymbol{V} \\ {}^{B}\boldsymbol{\omega} \times \boldsymbol{I}_Q\, {}^{B}\boldsymbol{\omega} \end{bmatrix} \tag{1}$$

where: ${}^{B}\boldsymbol{F}$ is the total force acting on the center of mass; ${}^{B}\boldsymbol{\tau}$ is the total torque; m_Q is the mass of the quadrotor; $\boldsymbol{I}_Q$ is the moment of inertia of the quadrotor; $\mathbb{I}_3$ is a 3×3 identity matrix; $\mathbb{O}_3$ is a 3×3 zero matrix; ${}^{B}\boldsymbol{V}$ is the linear velocity of the quadrotor in the body frame; and ${}^{B}\boldsymbol{\omega}$ is the angular velocity in the body frame.

$${}^{B}\boldsymbol{\omega} = [\,p \quad q \quad r\,]^T \tag{2}$$

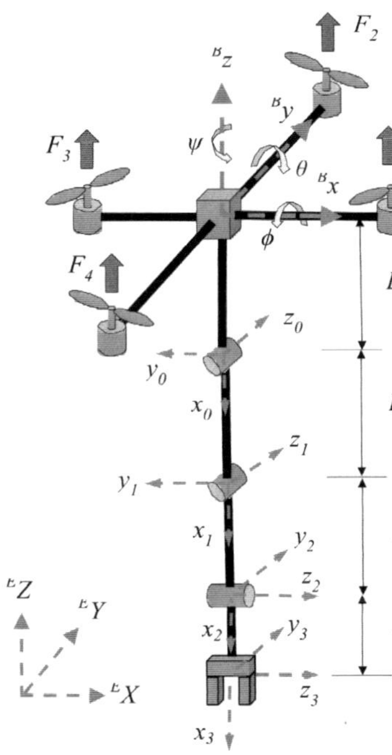

Figure 1. Quadrotor-Based Aerial Manipulator.

The controllable forces and moments of the quadrotor are thrust force U_1, yaw moment U_4, pitch moment U_3, roll moment U_2. The control inputs can be written as:

$$\begin{bmatrix} U_1 \\ U_2 \\ U_3 \\ U_4 \end{bmatrix} = \begin{bmatrix} \omega_1^2 + \omega_2^2 + \omega_3^2 + \omega_4^2 \\ \omega_4^2 - \omega_2^2 \\ \omega_3^2 - \omega_1^2 \\ \omega_4^2 - \omega_3^2 + \omega_2^2 - \omega_1^2 \end{bmatrix} \tag{3}$$

where ω_1, ω_2, ω_3, ω_4 are the rotor's angular velocities respectively.

Neglecting drag force, there are two forces acting on the quadrotor, the gravitational force and thrust force. The total torque is a result of the gyroscopic moments due to inertia of the rotors and thrust moments. In the body frame the gravitational force experienced by the quadrotor is:

$$^B\boldsymbol{F}_G = -m_Q g[-\sin\theta \quad \sin\phi\cos\theta \quad \cos\phi\cos\theta\,]^T \tag{4}$$

The thrust force experienced by the quadrotor in the body frame is:

$$^B\boldsymbol{F}_T = k[0 \quad 0 \quad U_1]^T \tag{5}$$

where k is the thrust factor. The thrust moment in the body frame is:

$$^B\boldsymbol{\tau}_T = [lkU_2 \quad lkU_3 \quad c_Q U_4]^T \tag{6}$$

where c_Q is a constant of proportionality; and l is the length of the quadrotor arms. The gyroscopic moment due to the rotating rotors is:

$$^B\boldsymbol{\tau}_G = I_r[q\bar{\omega}_r \quad -p\bar{\omega}_r \quad 0]^T \tag{7}$$

$$\bar{\omega}_r = -\omega_1 + \omega_2 - \omega_3 + \omega_4 \tag{8}$$

2.2. *Manipulator Model*

The Denavit-Hartenberg convention as found in [13] and [14] is used to describe the manipulator kinematics. A reference frame is attached to each link of the manipulator. The position of the origin of coordinate frame O_i, (with axis ($\boldsymbol{x}_i$, $\boldsymbol{y}_i$, z_i)), relative to coordinate frame O_{i-1}, (with axis ($\boldsymbol{x}_{i-1}$, $\boldsymbol{y}_{i-1}$, z_{i-1})), can be denoted by vector ${}^{i-1}\boldsymbol{p}_i$. The orientation of frame O_i relative to frame O_{i-1} can be denoted by the rotation matrix ${}^{i-1}\boldsymbol{R}_i$. The recursive Newton-Euler formulation as found in [13] is used in this paper to describe the manipulator dynamics.

2.2.1. Forwards recursion. In the Newton-Euler forward recursion the following are found [13]:

${}^{i}\boldsymbol{\omega}_i$ is the angular velocity of link i expressed in frame i:

$${}^{i}\boldsymbol{\omega}_i = {}^{i-1}\boldsymbol{R}_i^T\left({}^{i-1}\boldsymbol{\omega}_{i-1} + \dot{\theta}_i \boldsymbol{z}_0\right) \tag{9}$$

${}^{i}\dot{\boldsymbol{\omega}}_i$ is the angular acceleration of link i expressed in frame i:

$${}^{i}\dot{\boldsymbol{\omega}}_i = {}^{i-1}\boldsymbol{R}_i^T\left({}^{i-1}\dot{\boldsymbol{\omega}}_{i-1} + \ddot{\theta}_i \boldsymbol{z}_0 + \dot{\theta}_i\ {}^{i-1}\boldsymbol{\omega}_{i-1} \times \boldsymbol{z}_0\right) \tag{10}$$

${}^{i}\ddot{\boldsymbol{p}}_i$ is the linear acceleration of the origin of coordinate frame O_i relative to frame O_i:

$${}^{i}\ddot{\boldsymbol{p}}_i = {}^{i-1}\boldsymbol{R}_i^T\ {}^{i-1}\ddot{\boldsymbol{p}}_{i-1} + {}^{i}\dot{\boldsymbol{\omega}}_i \times {}^{i}\boldsymbol{r}_{i-1,i} + {}^{i}\boldsymbol{\omega}_i \times \left({}^{i}\boldsymbol{\omega}_i \times {}^{i}\boldsymbol{r}_{i-1,i}\right) \tag{11}$$

${}^{i}\ddot{\boldsymbol{p}}_{ci}$ is the linear acceleration of center of mass of link i relative to coordinate frame O_i:

$${}^{i}\ddot{\boldsymbol{p}}_{ci} = {}^{i}\ddot{\boldsymbol{p}}_i + {}^{i}\dot{\boldsymbol{\omega}}_i \times {}^{i}\boldsymbol{r}_{i,ci} + {}^{i}\boldsymbol{\omega}_i \times \left({}^{i}\boldsymbol{\omega}_i \times {}^{i}\boldsymbol{r}_{i,ci}\right) \tag{12}$$

*2.2.2. Backwards recursion.*In the backward recursion, the following are found [13]:

${}^{i}\boldsymbol{f}_i$ is the force exerted by link $i - 1$ on link i expressed in frame i:

$${}^{i}\boldsymbol{f}_i = {}^{i}\boldsymbol{R}_{i+1}\ {}^{i+1}\boldsymbol{f}_{i+1} + m_i\ {}^{i}\ddot{\boldsymbol{p}}_{ci} \tag{13}$$

${}^{i}\boldsymbol{\mu}_i$ is the moment exerted by link $i - 1$ on link i with respect to frame O_{i-1} expressed in frame i:

$${}^{i}\boldsymbol{\mu}_i = -{}^{i}\boldsymbol{f}_i \times \left({}^{i}\boldsymbol{r}_{i-1,i} + {}^{i}\boldsymbol{r}_{i,ci}\right) + {}^{i}\boldsymbol{R}_{i+1}\ {}^{i+1}\boldsymbol{\mu}_{i+1} + {}^{i}\boldsymbol{R}_{i+1}\ {}^{i+1}\boldsymbol{f}_{i+1} \times {}^{i}\boldsymbol{r}_{i,ci} + {}^{i}\boldsymbol{I}_i\ {}^{i}\dot{\boldsymbol{\omega}}_i + {}^{i}\boldsymbol{\omega}_i \times \left({}^{i}\boldsymbol{I}_i\ {}^{i}\boldsymbol{\omega}_i\right) \tag{14}$$

τ_i is the torque of link i:

$$\tau_i = {}^{i}\boldsymbol{\mu}_i^T\ {}^{i-1}\boldsymbol{R}_i^T\ \boldsymbol{z}_0 \tag{15}$$

2.3. Aerial Manipulator Model

Figure 1 shows the aerial manipulator consisting of a quadrotor, and a three-DoF manipulator (two-DoF arm and one-DoF wrist). Combining the quadrotor and the manipulator means that the manipulator no longer has a fixed base but has a floating base, that is the quadrotor. For a manipulator with a floating base, the initial linear velocity ${}^{0}\dot{\boldsymbol{p}}_0$, linear acceleration angular velocity ${}^{0}\ddot{\boldsymbol{p}}_0$, and angular acceleration ${}^{0}\dot{\boldsymbol{\omega}}_0$ are those of the quadrotor.

$${}^{0}\dot{\boldsymbol{p}}_0 = {}^{B}\boldsymbol{R}_0^T\ {}^{E}\boldsymbol{R}_B^T\ {}^{E}\dot{\boldsymbol{\Gamma}} \tag{16}$$

$${}^{0}\ddot{\boldsymbol{p}}_0 = {}^{B}\boldsymbol{R}_0^T\ {}^{E}\boldsymbol{R}_B^T\left({}^{E}\ddot{\boldsymbol{\Gamma}} - [0 \quad 0 \quad -g]^T\right) \tag{17}$$

$${}^{0}\boldsymbol{\omega}_0 = {}^{B}\boldsymbol{R}_0^T\ {}^{B}\boldsymbol{\omega} \tag{18}$$

$${}^{0}\dot{\boldsymbol{\omega}}_0 = {}^{B}\boldsymbol{R}_0^T\ {}^{B}\dot{\boldsymbol{\omega}} \tag{19}$$

where $^{E}\boldsymbol{\Gamma}$ is the position of the quadrotor in the inertial frame; and $^{B}\boldsymbol{R}_0$ is the rotation matrix from the manipulator base frame to the quadrotor body frame. The center of mass $\boldsymbol{C}_M$ of the aerial manipulator system in the body frame is:

$$\boldsymbol{C}_M = \frac{\boldsymbol{C}_{MQ}\, m_Q + \sum_{i=1}^{n} \boldsymbol{C}_{Mi}\, m_i}{m_Q + \sum_{i=1}^{n} m_i} \tag{20}$$

where $\boldsymbol{C}_{MQ}$ is the center of mass of the quadrotor expressed in the quadrotor body frame; and $\boldsymbol{C}_{Mi}$ is the center of mass of link i expressed in the quadrotor body frame. Expressing the moment of inertia of each link about the aerial manipulator center of mass in the quadrotor body frame:

$$\boldsymbol{I}_i = \boldsymbol{I}_{cm_i} + m_i \boldsymbol{S}(\boldsymbol{C}_i)\boldsymbol{S}(\boldsymbol{C}_i)^T \tag{21}$$

where $\boldsymbol{S}$ is the skew symmetric matrix of $\boldsymbol{C}_i$

$$\boldsymbol{C}_i = \boldsymbol{C}_{Mi} - \boldsymbol{C}_M \tag{22}$$

The moment of inertia $\boldsymbol{I}_S$ of the aerial manipulator is thus:

$$\boldsymbol{I}_S = \sum_{i=1}^{n} \boldsymbol{I}_i + \boldsymbol{I}_Q + m_Q \boldsymbol{S}(\boldsymbol{C}_Q)\boldsymbol{S}(\boldsymbol{C}_Q)^T \tag{23}$$

For the combined aerial manipulator system, the Newton-Euler formulation for the dynamics can be written as:

$$\begin{bmatrix} ^{B}\boldsymbol{F} \\ ^{B}\boldsymbol{\tau} \end{bmatrix} + \begin{bmatrix} ^{B}\boldsymbol{f}_1 \\ ^{B}\boldsymbol{\mu}_1 \end{bmatrix} = \begin{bmatrix} m_Q \mathbb{I}_3 & \mathbb{O}_3 \\ \mathbb{O}_3 & \boldsymbol{I}_S \end{bmatrix} \begin{bmatrix} ^{B}\dot{\boldsymbol{V}} \\ ^{B}\dot{\boldsymbol{\omega}} \end{bmatrix} + \begin{bmatrix} ^{B}\boldsymbol{\omega} \times m\, ^{B}\boldsymbol{V} \\ ^{B}\boldsymbol{\omega} \times \boldsymbol{I}_S\, ^{B}\boldsymbol{\omega} \end{bmatrix} \tag{24}$$

$$^{B}\boldsymbol{f}_1 = {^{B}\boldsymbol{R}_1}\, {^{1}\boldsymbol{f}_1} \tag{25}$$

$$^{B}\boldsymbol{\mu}_1 = {^{B}\boldsymbol{R}_1}\, {^{1}\boldsymbol{\mu}_1} \tag{26}$$

3. System Identification

Modelling and system identification of the aerial manipulator was performed in Matlab, with the full nonlinear model of the quadrotor, combined with the manipulator dynamics, being used for the simulations. The aerial manipulator was simulated with the quadrotor altitude and attitude controlled using Proportional-Derivative (PD) control. The quadrotor was controlled to try maintain a constant Z height of 1.5 m, while trying to keep the roll ϕ, pitch θ and yaw ψ angles at 0 rad. The manipulator joint angles were set to perform trapezoidal velocity profiles resulting in the joint angle trajectories shown in figure 2a. For validation the manipulator angle trajectories shown in figure 2b were used. The quadrotor linear accelerations $\ddot{X}$, $\ddot{Y}$, $\ddot{Z}$ and angular accelerations $\ddot{\phi}$, $\ddot{\theta}$, $\ddot{\psi}$ in the inertial frame were then identified given the control inputs U_1 to U_4 and the manipulator joint angular positions, velocities and accelerations.

First, Auto Regressive with eXogenous inputs (ARX) models of orders one to thirty were attempted for each of the linear and angular accelerations. If ARX models were not successful, then ARMAX models of orders one to thirty and nonlinear ARX models were also attempted. For the nonlinear ARX models, both wavelet and sigmoid networks nonlinearity estimators were attempted.

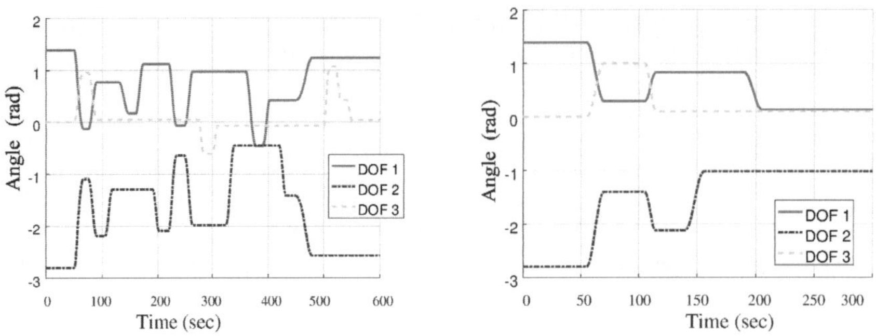

(a) Estimation (b) Validation

Figure 2. Manipulator joint angle trajectory for system identification validation.

3.1. *Results*

For the linear acceleration $\ddot{X}$ a third-order ARX model was identified as shown in figure 3. Both validation and estimation data had over 99% fit.

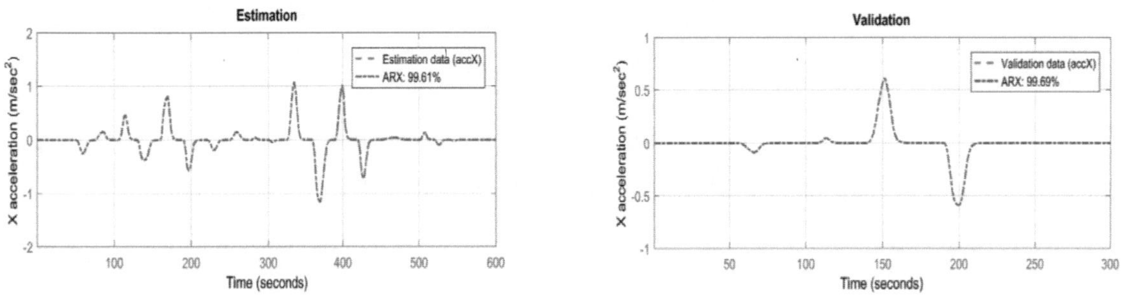

Figure 3. System Identification Results - X acceleration.

Similarly, for the linear acceleration $\ddot{Y}$ a third-order ARX model was identified as shown in figure 4. As with $\ddot{X}$ both validation and estimation data for $\ddot{Y}$ had over 99% fit.

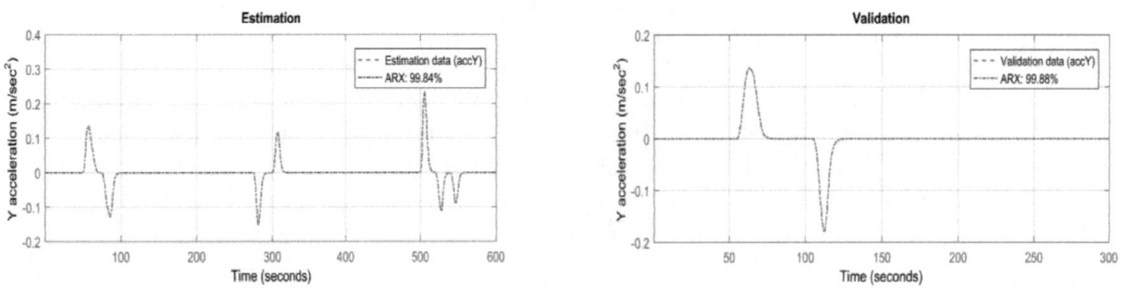

Figure 4. System Identification Results - Y acceleration.

The linear acceleration $\ddot{Z}$ had poor results for ARX models. Nonlinear ARX models were also investigated, however they did not significantly improve on the results of the linear ARX models. A fifth-order ARMAX was identified as shown in figure 5. In this case, estimation data had a 69.5% fit and validation had a 62% fit. It should be noted that the $\ddot{Z}$ linear acceleration had, overall, a significantly smaller magnitude than that of the linear accelerations $\ddot{X}$ and $\ddot{Y}$.

For roll angular acceleration $\ddot{\phi}$, a third-order ARMAX model was identified with estimation and validation fits of over 79% as shown in figure 6. A third-order ARX model was also identified with estimation and validation fits of over 78.8%.

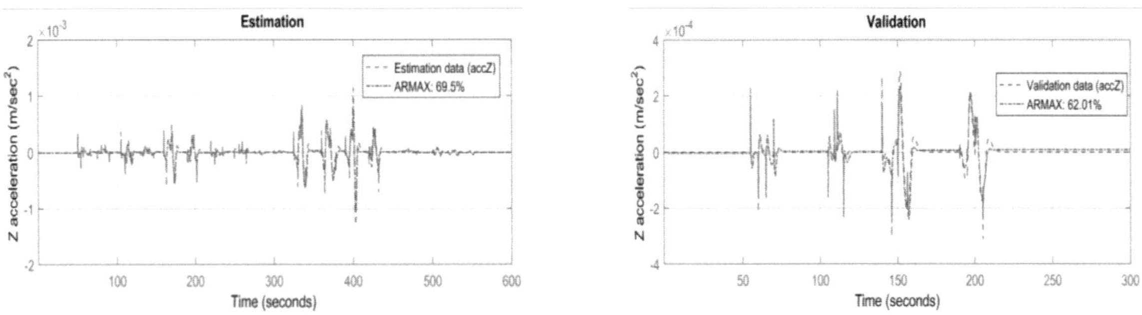

Figure 5. System Identification Results - Z acceleration.

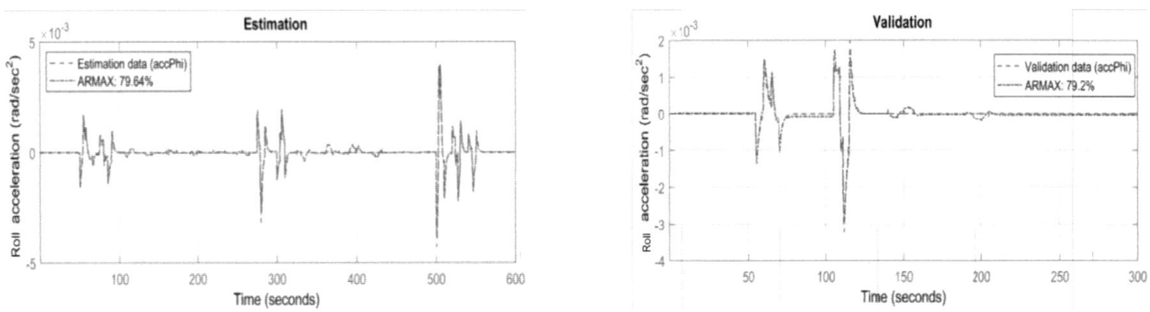

Figure 6. System Identification Results - Roll acceleration.

For pitch angular acceleration $\ddot{\theta}$, a sixth-order ARMAX model was identified with an estimation fit of 87.5% and validation fit of 84.9% as shown in figure 7.

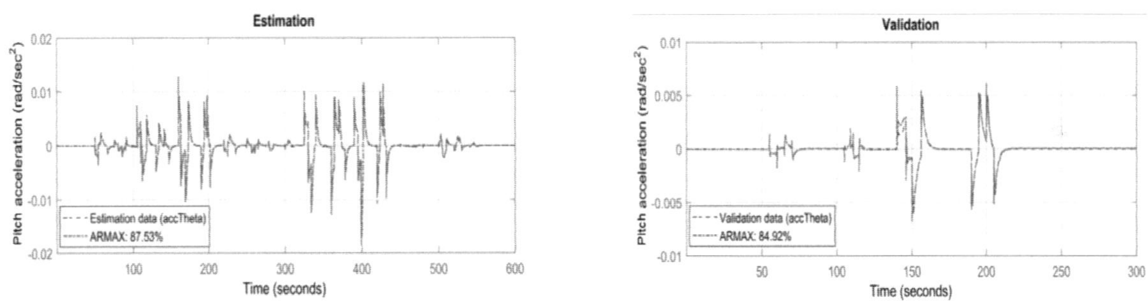

Figure 7. System Identification Results - Pitch acceleration.

For yaw angular acceleration $\ddot{\psi}$ a third-order ARX model resulted in a fit of 88.3% estimation fit and an 86.1% validation fit as shown in figure 8.

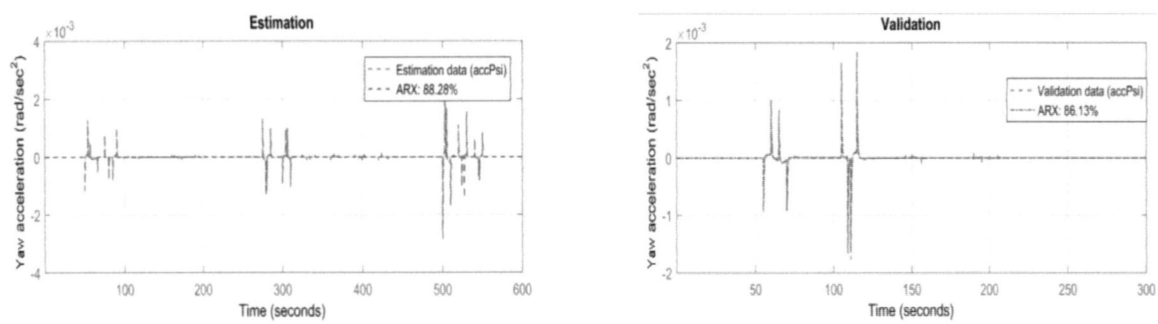

Figure 8. System Identification Results - Yaw acceleration.

4. Conclusions and Future Work

This paper presents the system identification of a simulated quadrotor-based aerial manipulator. The aerial manipulator dynamics model for a quadrotor with a two-link manipulator with a one-DoF wrist was successfully identified for the case of the quadrotor under altitude and attitude control with the manipulator in motion. ARX models were identified for linear accelerations $\ddot{X}$ and $\ddot{Y}$ and yaw angular acceleration $\ddot{\psi}$. For linear acceleration $\ddot{Z}$ and angular accelerations $\ddot{\phi}$ and $\ddot{\theta}$, ARMAX models were identified.

Future work includes performing system identification from actual flight data of a quadrotor-based aerial manipulator. While satisfactory results were obtained for linear acceleration $\ddot{Z}$, non-linear ARMAX models could also be investigated for system identification of the aerial manipulator to improve the identification results of $\ddot{Z}$.

References

[1] X. Zhang, X. Li, K. Wang, and Y. Lu, "A survey of modelling and identification of quadrotor robot," *Abstract and Applied Analysis*, vol. 2014, no. Article ID 320526, pp. 1–16, 2014.

[2] L. Derafa, T. Madani, and A. Benallegue, "Dynamic modelling and experimental identification of four rotors helicopter parameters," in *Proceedings of the IEEE International Conference on Industrial Technology*, (Mumbai), pp. 1834–1839, 15–17 Dec 2006.

[3] W. Dong, G.-Y. Y. Gu, X. Zhu, and H. Ding, "Development of a quadrotor test bed - Modelling, parameter identification, controller design and trajectory generation," *International Journal of Advanced Robotic Systems*, vol. 12, no. 7, pp. 1–14, 2015.

[4] M. Elsamanty, A. Khalifa, M. Fanni, A. Ramadan, and A. Abo-Ismail, "Methodology for identifying quadrotor parameters, attitude estimation and control," in *2013 IEEE/ASME International Conference on Advanced Intelligent Mechatronics: Mechatronics for Human Wellbeing, AIM 2013*, (Wollongong, NSW), pp. 1343–1348, 9–12 Jul 2013.

[5] O. Araar, M. Z. Mimouni, K. Fellah, and H. Osmani, "Identification and Control of a Multirotor UAV in the Presence of Actuator Asymmetry," in *25th Mediterranean Conference on Control and Automation (MED)*, (Valletta), pp. 1035–1040, 3–6 Jul 2017.

[6] K. S. Hatamleh, M. Al-Shabi, A. Al-Ghasem, and A. A. Asad, "Unmanned Aerial Vehicles parameter estimation using Artificial Neural Networks and Iterative Bi-Section Shooting method," *Applied Soft Computing*, vol. 36, no. Nov 2015, pp. 457–467, 2015.

[7] G. Lee, D. Y. Jeong, N. D. Khoi, and T. Kang, "Attitude Control System Design for a Quadrotor Flying Robot," in *8th International Conference on Ubiquitous Robots and Ambient Intelligence (URAI)*, (Incheon), pp. 74–78, 23-26 Nov 2011.

[8] I. Stanculeanu and T. Borangiu, "Quadrotor Black-Box System Identification," *International Journal of Mechanical, Aerospace, Industrial, Mechatronic and Manufacturing Engineering*, vol. 5, no. 6, pp. 1025– 1028, 2011.

[9] R. J. A. Schreurs, S. Weiland, H. Tao, Q. Zhang, J. Zhu, Y. Zhu, and C. Xu, "Open loop system identification for a quadrotor helicopter system," in *IEEE International Conference on Control and Automation, ICCA*, (Hangzhou), pp. 1702–1707, 12–14 Jun 2013.

[10] M. Bergamasco and M. Lovera, "Identification of Linear Models for the Dynamics of a Hovering Quadrotor," *IEEE Transactions on Control Systems Technology*, vol. 22, no. 5, pp. 161–166, 2014.

[11] J. O. Pedro and A. J. Crouse, "Direct Adaptive Neural Control of a Quadrotor Unmanned Aerial Vehicle," in *10th Asian Control Conference*, (Kota Kinabalu), pp. 1–6, 31 May–3 Jun 2015.

[12] J. O. Pedro and M. Dangor, "Black-Box System Identification of a Quadrotor UAV Using Dynamic Neural Network," in *CEAS Specialist Conference on Guidance, Navigation and Control*, (Warsaw), pp. 1–18, 25–27 Apr 2017.

[13] L. Sciavicco and B. Siciliano, *Modelling and Control of Robot Manipulators*. London: Springer-Verlag, 2005.

[14] K. Waldron and J. Schmiedeler, *Springer Handbook of Robotics*, ch. Kinematics, pp. 9–34. Springer-Verlag, 2008.

IOP Conf. Series: Journal of Physics: Conf. Series **1016** (2018) 012007 doi:10.1088/1742-6596/1016/1/012007

Acknowledgement
This work is based on the research supported in part by the National Research Foundation of South Africa for the grant, Unique Grant No. 107025.

IOP Conf. Series: Journal of Physics: Conf. Series **1016** (2018) 012008 doi:10.1088/1742-6596/1016/1/012008

Trajectory tracking control for a nonholonomic mobile robot under ROS

Khadir Lakhdar Besseghieur[1], Radosław Trębiński[1], Wojciech Kaczmarek[1], Jarosław Panasiuk[1]

[1] Faculty of Mechatronics and Aerospace, Military University of Technology, Kaliskiego 2, 00-908, Warsaw, Poland

Corresponding author: besseghieurkh@hotmail.fr

Abstract. In this paper, the implementation of the trajectory tracking control strategy on a ROS-based mobile robot is considered. Our test-bench is the nonholonomic mobile robot 'TURTLEBOT'. ROS facilitates considerably setting-up a suitable environment to test the designed controller. Our aim is to develop a framework using ROS concepts so that a trajectory tracking controller can be implemented on any ROS-enabled mobile robot. Practical experiments with 'TURTLEBOT' are conducted to assess the framework reliability.

1. Introduction

Recently, a lot of research have been conducted in order to control mobile robots [1], [2] and [3]. The mobile robots with a steering wheel (unicycle) or two independent drive wheels like 'TURTLEBOT' are examples with considerably important engineering interest. A big part of mobile robots are subjected to non-integrable kinematic constraints which cannot be reduced to holonomic ones, they are called non-holonomic constraints. The problem known as the tracking trajectory problem is when the reference point on the robot is required to follow the desired trajectory. In other words, the distance between the reference point on the robot and the current path reference point must be kept as small as possible. Controlling a mobile robot to track a specific trajectory has several practical applications particularly while operating in known map environments. Robots running on ROS like TURTLEBOT are commonly used to autonomously navigate in known environments. The autonomous navigation algorithm must contain a trajectory planner which generates the appropriate trajectory with the purpose of arriving at a particular location, patrolling through specified area and at the same time avoiding collisions with different kinds of obstacles. Once the trajectory is generated, the robot must actually track it. Therefore, the trajectory tracking control plays a substantial role in the autonomous navigation algorithm. The aim of this work is to develop a ROS-based framework that allows researchers to implement their trajectory tracking control algorithms on ROS-enabled mobile robots.

A number of trajectory tracking control laws for nonholonomic systems can be found in the literature and different approaches have been used to tackle this problem. Among the first approaches, Lyapunov stable time-varying state-tracking control laws were pioneered in [4], [5] and [6]. A model predictive controller is proposed in [7]. To the best of our knowledge, the trajectory tracking control implementation on a ROS-enabled robot is not considered in the literature. In this paper, we propose a new framework based on ROS concepts for all the ROS-enabled robots so that real experiments for the trajectory tracking control problem can be conducted effectively. The reminder of this paper is organized as follows: the next section is dedicated for introducing 'TURTLEBOT' and its kinematic model, then the tracking problem is formulated and it is concluded by the controller design. In Section 3, the proposed ROS-framework is detailed. Experimental results are presented and discussed in section 4. Finally, we conclude this paper and list some future work directions in section 5.

2. Tracking problem formulation for 'TURTLEBOT'

TURTLEBOT is a highly capable autonomous mobile platform for developing robot applications. The hardware structure can be divided into three main parts: A KOBUKI base, a netbook computer, and sensors. The Kobuki base is a differential drive mobile robot base with two passive caster wheels for balancing. It is modeled as a two-wheeled differential drive robot. Our netbook runs Ubuntu 14.04 LTS OS and has ROS Indigo running on it. These robots are considered as a class of computer-controlled vehicles whose motion can be described or transformed into the following model of constrained movement in a plane as in equation (1). With *(x; y)* represent the planar coordinates, θ is the orientation with respect to the X-axis, v and w denote the translational and rotational velocity respectively:

$$\begin{pmatrix} \dot{x} \\ \dot{y} \\ \dot{\theta} \end{pmatrix} = \begin{bmatrix} \cos\theta & 0 \\ \sin\theta & 0 \\ 0 & 1 \end{bmatrix} \begin{pmatrix} v \\ w \end{pmatrix} \tag{1}$$

It is easy to see that the TURTLEBOT is governed by the following nonholonomic constraint:

$$\dot{x} \sin\theta - \dot{y} \cos\theta = 0 \tag{2}$$

Now consider a reference trajectory of the form:

$$\begin{pmatrix} \dot{x_r} \\ \dot{y_r} \\ \dot{\theta_r} \end{pmatrix} = \begin{bmatrix} \cos\theta_r & 0 \\ \sin\theta_r & 0 \\ 0 & 1 \end{bmatrix} \begin{pmatrix} v_r \\ w_r \end{pmatrix} \tag{3}$$

Where v_r and w_r are the linear and angular reference velocity respectively. The tracking errors and the errors dynamics can be defined as in [4] by:

$$\begin{pmatrix} e_x \\ e_y \\ e_z \end{pmatrix} = \begin{bmatrix} \cos\theta & \sin\theta & 0 \\ -\sin\theta & \cos\theta & 0 \\ 0 & 0 & 1 \end{bmatrix} \begin{pmatrix} x_r - x \\ y_r - y \\ \theta_r - \theta \end{pmatrix} \tag{4}$$

$$\begin{pmatrix} \dot{e_x} \\ \dot{e_y} \\ \dot{e_\theta} \end{pmatrix} = \begin{bmatrix} e_y w - v + v_r \cos e_\theta \\ -e_x w + v_r \sin e_\theta \\ w_r - w \end{bmatrix} \tag{5}$$

To globally asymptotically stabilize the error dynamics in equation (5), JIANG and NIJMEIJER proposed their approach in [6], where time varying state feedback control laws based on the backstepping technique are proposed. Using the Lyapunov's direct method, global controller is derived for the trajectory tracking problem. Referring to [6], if $\dot{v_r}, \dot{w_r}, v_r$ and w_r are all bounded and with the condition of either v_r or w_r does not converge to zero, the error dynamics in equation (5) are globally asymptotically stabilized using the control inputs in equation (6). The proof of convergence are omitted here, the reader is referred to [6]. With $a=(w_r^2 + b v_r^2)^{1/2}, \varepsilon > 0$ and $b>0$, the control laws are:

$$\begin{pmatrix} v \\ w \end{pmatrix} = \begin{pmatrix} v_r \cos e_\theta + K_x e_x \\ w_r + (K_y e_y \sin(e_\theta)/ e_\theta) + K_\theta . e_\theta \end{pmatrix} \text{ with } \begin{pmatrix} K_x \\ K_y \\ K_\theta \end{pmatrix} = \begin{pmatrix} 2\varepsilon\, a \\ b\,|v_r| \\ 2\varepsilon\, a \end{pmatrix} \tag{6}$$

3. Tracking controller ROS implementation

The main contribution of this paper, mainly the proposed ROS framework is detailed in this section, the ROS packages and the nodes that are relevant in our framework are also presented. ROS is a Linux-based, an open source software package that provides a software framework to aid in the development of complex robotic applications [8]. The proposed framework needs first an offline-built 2D map of the environment necessary for the robot's localization. The map is generated using the open-source package 'G-Mapping' where the robot is teleoperated around the environment; the system uses the point cloud generated by Kinect and the Odometry information from the encoders to build a 2D map and stores it locally on the robot's laptop. Our framework mainly consists of: the localization packages and the control node. The two parts are connected through the topic named "*/tf*".

3.1. Localization

To track a desired path, it is crucial that the robot should be aware of where it is itself as well as where the rest of the world is in relation to itself. The ROS navigation stack is used to provide the localization of the robot based on the Adaptive Monte-Carlo Localization (AMCL) approach presented in [9]. It is based on a weighted particle system in which each particle represents an estimated pose of the robot and consists of two phases of calculation: the prediction and update phases. AMCL combines the on-board encoders' measurements and the data provided by the Kinect sensor to provide an accurate estimation of the robot position. The framework needs the 'map_server' to upload the offline-built map which is necessary for the AMCL node to estimate the robot's coordinates vis-à-vis the map's origin. AMCL then publishes the transform between the two frames '/map' and '/base_footprint' into the topic 'tf'. The first frame corresponds to the map's origin whereas the latter corresponds to the robot's base center. In other words, the robot's coordinates in the map's coordinate system are published into the topic 'tf'. To read this topic's data, we opted for the tf library [10]. It was designed to provide a standard way to keep track of coordinate frames and transform data within an entire system such that individual component users can be confident that the data is in the coordinate frame that they want without requiring knowledge of all the coordinate frames in the system. The tf library provides a listener module which allows us to track the robot's coordinates in the absolute frame named '/map'.

3.2. Control node

The control node code was developed in C++. The programmed control laws need as entries the robot's coordinates provided by the localization nodes. The control loop with period T=10ms starts with a 'tf_listener' from the 'tf' library [10] is used to obtain the frame transform between the frames '/map' and '/base-footprint' published in the topic 'tf'. Having obtained the actual robot's position, the trajectory tracking errors are evaluated and then the input control values are computed. A publisher is programmed to publish the obtained values to the corresponding topics in order to control the robot's actuators. To respect the physical limitations of the robot, the control speeds are limited within the control node program before published. A saturation of the command velocities [1] that preserves the current curvature $k = w/v$ is performed as:

$$\begin{cases} \sigma = max\{|v|/v_{max}, |w|/w_{max}, 1\} \\ v_c = sign(v)v_{max}, w_c = w/\sigma \; if \; \sigma = |v|/v_{max} \\ w_c = sign(w)w_{max}, v_c = v/\sigma \; if \; \sigma = |w|/w_{max} \\ v_c = v, w_c = w \; if \; \sigma = 1 \end{cases} \quad (7)$$

All the aforementioned nodes necessary for our framework are executed within one launch file, where all the parameters needed by the nodes are specified as well. The custom launch file is executed with one line command from a terminal.

4. Experimental results

To test the proposed framework, two reference trajectories are selected to be tracked by the robot. The mathematical model of the desired trajectories: circular and 8-shape trajectory are presented in equation (8.a) and (8.b) respectively. $(\boldsymbol{x_c}, \boldsymbol{y_c})$ represent the center coordinates and R is the radius of the circle, k_1, $k_2>0$, $k_1 \neq k_2$ and $k_3=0, 1$. In experimentation, the robot starts from the map's origin i.e. with an initial state error from the reference trajectory. Note that in both experiments the maximum control velocities were limited, as explained in section 3.2 and the maximum allowed linear and angular velocity in both simulation and experiments were $\boldsymbol{v_{max} = 0.4\, m/s, w_{max} = 0.8\, rd/s}$. The obtained data are recorded and saved on the robot's laptop then drawn using MATLAB.

IOP Conf. Series: Journal of Physics: Conf. Series **1016** (2018) 012008 doi:10.1088/1742-6596/1016/1/012008

$$\begin{cases} x_r = x_c + R.\sin(k.t) \\ y_r = y_c - R.\cos(k.t) \\ \theta_r = k.t \\ v_r = k.R \\ w_r = k \end{cases} \quad (a), \qquad \begin{cases} x_r = x_c + R.\sin(k_1.t) \\ y_r = y_c + R.\sin(k_2.t) \\ \theta_r = Atan2(\dot{y_r}(t), \dot{x_r}(t)) + k_3\pi \\ v_r = \pm\sqrt{\dot{y_r}^2(t) + \dot{x_r}^2(t)} \\ w_r = \dfrac{\ddot{y_r}(t)\,\dot{x_r}(t) - \ddot{x_r}(t)\,\dot{y_r}(t)}{\dot{y_r}^2(t) + \dot{x_r}^2(t)} \end{cases} \quad (b) \qquad (8)$$

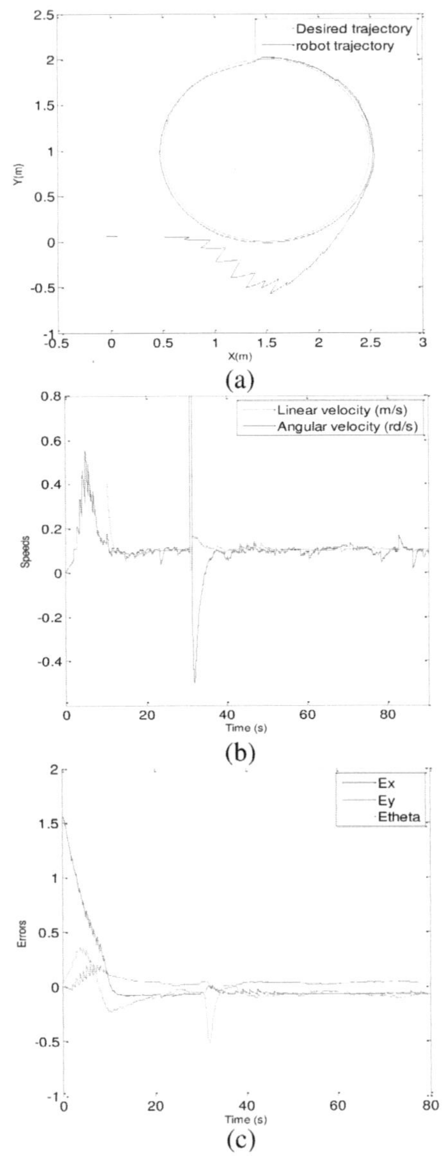

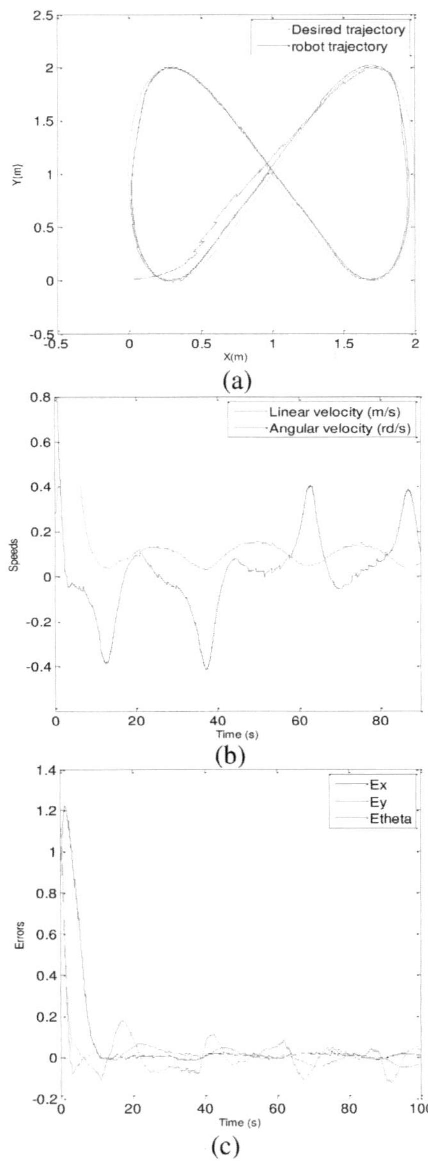

Figure 1. circular trajectory tracking. **Figure 2.** 8-shape trajectory tracking.

Before implementing the controller on the real robot, a simulation was conducted first to assess the effectiveness of the trajectory tracking controller in [6] before evaluating the framework itself. The environment 'GAZEBO' was used for simulating a virtual 'TURTLEBOT' seeking to track a circular trajectory. For the real tests on Turtlebot, the robot is initially at the origin (0,0) and required first to drive along a circle centered in $(x_c, y_c)=(1.5,1)$ with k=0.1 and $R=1$. Figure 1.a shows that the robot tracks the desired trajectory with a negligible error. The tracking errors in figure 1.c converge after approximately 15 seconds to some negligible values around zero. Control inputs for the real robot in figure 1.b finish by converging around the desired trajectory velocities (v_r=0.1 m/s, w_r=0.1 m/s) as obtained by (8). In figure 2, the robot is required to track the 8-shape trajectory where $(x_c,y_c)=(1,1)$, $k_1=2\pi/100$ and $k_2=2*k_1$. This a more challenging task since the desired trajectory velocities are time variant in this case. Figures 2.a and 2.c show that the robot tracks the desired trajectory with negligible errors. The control inputs are illustrated in figure 2.b and are time variant but they lie within the limited velocity ranges. We can observe in all graphs that the errors do not converge to the absolute zero. This is due only to the localization error where the position estimate delivered by the AMCL node is not very accurate with using only Kinect. This can be improved with incorporating a laser scanner.

5. Conclusion

In this work, a ROS-framework is proposed to implement a trajectory tracking controller on a ROS-enabled mobile robot. Experiments were conducted on a 'TURTLEBOT' by implementing the controller proposed in [6]. Two different trajectories were successfully tracked by the robot which confirms the reliability of the proposed framework. Future works lie on implementing under ROS more complicated and sophisticated control laws as well as improving the robot's self localization. A more developed framework for controlling several mobile robots simultaneously under ROS is under study.

References

[1] Klančar G and Škrjanc I 2007 Tracking-error model-based predictive control for mobile robots in real time *Robotics and Autonomous Systems* 460-469

[2] Z. Li, J. Deng and R. Lu 2016 Trajectory-Tracking Control of Mobile Robot Systems Incorporating Neural-Dynamic Optimized Model Predictive Approach. *IEEE Transactions on Systems, Man, and Cybernetics: Systems* 740-749

[3] Mobayen S 2015 Fast terminal sliding mode tracking of nonholonomic systems with exponential decay rate *IET Control Theory & Applications* 1294–1301

[4] Kanayama Y., Kimura Y., Miyazaki F, Noguchi T 1990 A stable tracking control method for an autonomous mobile robot. *IEEE International Conf. on Robotics and Automation* 384-389

[5] Samson C., Ait-abderrahim K 1991 Feedback control of a nonholonomic wheeled cart in Cartesian space. *1991 IEEE International Conference on Robotics and Automation* 1136–1141

[6] Z-P JIANGdagger, H.NIJMEIJER 1997 Tracking Control of Mobile Robots: A Case Study in Backstepping. *Automatica* Volume 33, Issue 7 1393-1399

[7] Ollero A. and Amidi O 1991 Predictive path tracking of mobile robots. *Proceedings of 5th Int.Conf. on Advanced Robotics, Robots in Unstructured Environments* 1081 – 1086

[8] Quigley M., Gerky B., and Smart W. D 2015 Programming Robots with ROS, a Practical Introduction to the Robot Operating System *First. O'Reilly Media, Inc*

[9] Hennes D, Claes D, Meeussen W, 2012 Multi-robot collision avoidance with localization uncertainty. *11th Int. Conf. on Autonomous Agents and Multiagent Systems* 147-154.

[10] T. Foote 2013 tf: The transform library *In Technologies for Practical Robot Applications (TePRA), 2013 IEEE International Conference on*, 1–6.

IOP Conf. Series: Journal of Physics: Conf. Series **1016** (2018) 012009 doi:10.1088/1742-6596/1016/1/012009

Output Tracking of Some Classes of Non-minimum Phase Nonlinear Systems By Redefinition Output Approach

Janson Naiborhu[1], **Firman**[2], **and Edwin Setiawan**[3]

[1,3]Industrial & Financial Mathematics Group, Faculty of Mathematics and Natural Science, Institut Teknologi Bandung. Gedung CAS, Lantai 4, Jln. Ganesa 10, Bandung 40132, Indonesia

[2]Department of Mathematics Faculty of Mathematics and Natural Science, Universitas Hasanuddin, Makasar, Indonesia.

E-mail: [1]`janson@math.itb.ac.id`,[2]`firman.math11@gmail.com`, [3]`edwin.setiawan.n@gmail.com`

Abstract. In this paper, we present the redefinition output approach for output tracking of some classes of non-minimum phase nonlinear systems without and with uncertainty. Input-Output linearization and gradient descent methods are applied to design the control. The good result of this approach is demonstrated by 3 examples.

1. Introduction

A system is called non-minimum phase if a nonlinear state feedback can hold the system output identically zero while the internal dynamics becomes unstable [1]. Output tracking problem for nonlinear non-minimum phase systems is a rather difficult issue in control theory. Most of researcher restrict their research to a special class nonlinear system only. The input-output linearization is one of the most available methods [1] for minimum phase besides the modified gradient descent control [2]. In this paper, both methods are applied to design control after non-minimum phase nonlinear system is transformed to minimum phase by redefinition output of the system. Results on stabilization of non-minimum phase system in the output feedback form have been presented in [3], [4], [5]. The main idea in [3], [4], [5] is output reconstruction such that the original nonlinear systems becomes minimum phase with respect to a new output. Results on output tracking of some classes of non-minimum phase nonlinear system have been presented in [6], [7]. In [6], The design of the input control is based on the exact linearization.

In this paper we give 3 examples to demonstrate how to track the output of the system by redefinition output combine with input-output linearization or modified gradient descent control. First example, the nonlinear system is assumed exact linearizable, the second example, the nonlinear system has the relative degree is $n - 1$, n is the dimension of the system. The third example, the nonlinear system has the relative degree is $n-1$ with uncertainty parameter. For relative dgree is 1, Dimitar Ho and J.Karl Hedrick has done in [8].

2. Problem Statement

Consider the following SISO affine nonlinear control system

$$\dot{\mathbf{x}} = \mathbf{f}(\mathbf{x}, \theta) + \mathbf{g}(\mathbf{x}, \theta)u, \tag{1}$$

$$y = h(\mathbf{x}) \tag{2}$$

where $\mathbf{x} \in \mathcal{R}^n$ is the state vector, $u \in \mathcal{R}$ is the control input, $y \in \mathcal{R}$ is the measured output, and θ is the parameter uncertainty. $\mathbf{f} : \mathcal{R}^n \to \mathcal{R}^n$ is a smooth function with $\mathbf{f}(0) = 0$, $\mathbf{g} : \mathcal{R}^n \to \mathcal{R}^n$ and $h : \mathcal{R}^n \to \mathcal{R}$ are smooth functions. Assume also that $h(0) = 0$. If the nonlinear system (1)-(2) has relative degree r, $(r < n)$ at x°, by input-output linearization [1], the system (1)-(2) can be transformed to

$$S = \begin{cases} \sum_{ext} : \begin{cases} \dot{\xi}_k = \xi_{k+1}, \quad k = 1, \cdots, r-1 \\ \dot{\xi}_r = a(\xi, \eta, \theta) + b(\xi, \eta, \theta)u \end{cases} \\ \sum_{int} : \dot{\eta} = q(\xi, \eta, \theta) \\ y = \xi_1 = h(\mathbf{x}), \end{cases} \tag{3}$$

with the internal dynamics

$$\sum_{int} : \dot{\eta} = q(\xi, \eta, \theta). \tag{4}$$

The stability of the internal state η is required to guarantee the output system $y(t)$ tracks the desired output $y_d(t)$. Our objective is to design input such that the output $y(t)$ tracks the desired output $y_d(t)$ while keeping the state bounded.

For case $b(\xi, \eta, \theta) \neq 0$ for $t \geq 0$, we apply input-output linearization method, i.e.,

$$u_r = \frac{1}{b(\mathbf{z})} \left(-a(\mathbf{z}) + v\right), \tag{5}$$

where $\mathbf{z} = [\xi_1, \cdots \xi_r, \eta_1, \cdots, \eta_{n-r}]$, $v = c_0 z_1 + c_1 \dot{z}_2 + \cdot + c_n z_1^{(n)}$ and the value of $c_i; i = 0, \cdots n$ is chosen such that the real part of the eigen values of polynomial $p(s)$

$$p(s) = c_n s^n + c_{n-1} s^{r-1} + ... + c_1 s^1 + c_o$$

are negative, $z_1 = h(\mathbf{x})$.

For case $b(\xi, \eta, \theta) = 0$ for a time t, we apply the modified gradient descent control[7]. Let $\eta(t)$ is a virtual output of the systems and $\eta_d(t)$ is the virtual desired output,and equilibrium point of the internal dynamics of normal form of the system. Then we find r_η as relative degree of the system if $\eta(t)$ is the output of the system. We know that $\eta \in \mathcal{R}^{n-r}$ then $r_\eta = [r_\eta^1, \cdots, r_\eta^{n-r}]$.

Based on $y(t)$, $\eta(t)$ and their derivatives, we construct the performance index as a descent function as follows,

$$F_0(y(t), \eta(t)) = \left(\sum_{j=0}^{r} a_j (y^{(j)}(t) - y_d^{(j)}(t)) \right)^2 + \sum_{i=1}^{n-r} \left(\sum_{j=0}^{r_\eta^i} b_j^i (\eta_i^{(j)}(t) - \eta_{di}^{(j)}(t)) \right)^2, \tag{6}$$

where the constants $a_0, \cdots, a_r; b_0^i, \cdots, b_{r_\eta^i}^i$, $i = 1, \cdots, n-r$ will be chosen such that eigenvalues of polynomials

$$a_r s^r + a_{r-1} s^{r-1} + \cdots + a_1 s + a_0, \tag{7}$$

$$b_{r_\eta^i}^i s^{r_\eta^i} + b_{r_\eta^i - 1}^i s^{r_\eta^i - 1} + \cdots + b_1^i s + b_0^i, \tag{8}$$

are real negative. Defines the descent function F_0 as a quadratic function to ensure that the function F_0 has a minimum value.

The modified gradient descent control is

$$\dot{u} = -\frac{dF_0}{du} + v, \tag{9}$$

where v is an artificial input,

$$v = \begin{cases} k(\mathbf{x}, u) & if \quad \frac{\partial F_0}{\partial u} \neq 0 \\ 0 & if \quad \frac{\partial F_0}{\partial u} = 0, \end{cases} \tag{10}$$

with

$$k(\mathbf{x}, u) = \frac{1}{\frac{\partial F_0}{\partial u}} \left(-\frac{\partial F_0}{\partial x}\dot{x} - \sqrt{\left(\frac{\partial F_0}{\partial x}\dot{x}\right)^2 + \left(\frac{\partial F_0}{\partial u}\right)^2} \right). \tag{11}$$

3. Redefinition Output Approach

3.1. Exact Linearization

Consider the following SISO affine nonlinear control system

$$\begin{aligned} \dot{x_1} &= x_2 + 2x_1^2 \\ \dot{x_2} &= x_3 + u \\ \dot{x_3} &= x_1 + x_3 \end{aligned} \tag{12}$$

$$y = x_1; \quad y_d(t) = \sin t. \tag{13}$$

By using the output $\lambda(x) = x_3$, the nonlinear system (12) can be linearized exactly.

$$\begin{aligned} \dot{z}_1 &= z_2 \\ \dot{z}_2 &= z_3 \\ \dot{z}_3 &= a(\mathbf{z}) + u, \end{aligned} \tag{14}$$

where $a(\mathbf{z}) = z_1 + z_2 + (2(z_2 - z_1) + 1)(z_3 - z_2 - 2(z_2 - z_1)^2 + 2(z_2 - z_1)^2)$. By input-output linearization technique we get

$$u = -a(\mathbf{z}) + v. \tag{15}$$

Let $y_d(t) = \sin(t) = x_{1d}(t)$. Next, we choose $z_{1d}(t)$ such that if $z_1(t)$ tracks $z_{1d}(t)$ then $y(t)$ tracks the desired output $y_d(t)$. Consider the equation : $\dot{x}_3 = x_1 + x_3$. By replacing x_1 with $x_{1d}(t) = \sin(t)$, we have a differential equation $\dot{x}_3 - x_3 = sin(t)$. Then, we solve the differential equation to obtain $x_3 = 1/2(-sin(t) - cos(t))$. This solution we state as $x_{3d}(t) = 1/2(-sin(t) - cos(t))$. Thus, for the output tracking problem we have

$$v = \frac{1}{a_3} \dot{z}_{3d} - \sum_{i=1}^{3} a_{i-1}(z_i - z_{id}). \tag{16}$$

The simulation results is given in Figure 1.

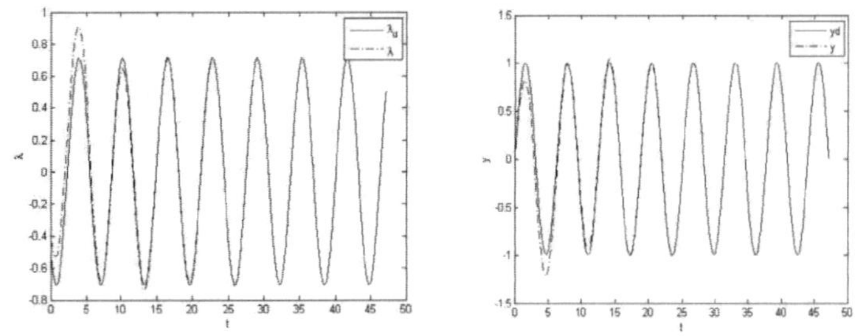

Figure 1. Left : Output Tracking (exact linearization), x_3 *to* x_{3d}; Right: Output Tracking (original system) y *to* y_d

3.2. Relative degree $r = n - 1$

Consider the nonlinear system (SISO)

$$\begin{aligned} \dot{x_1} &= x_2 \\ \dot{x_2} &= x_3 + x_1x_3 \\ \dot{x_3} &= x_4 - u + x_1x_3 \\ \dot{x_4} &= u - 2x_1x_3 \end{aligned} \qquad (17)$$

$$y = x_1. \qquad (18)$$

The nonlinear system (17)-(18) has relative degree 3 at any point x_0 (relative degree of the system is not well defined). Because the stability of zero dynamic is unstable, the nonlinear system (17)-(18) is the non-minimum phase. Now, redefining output $z_1 = x_1 + 2x_2 + 2x_3 + 2x_4$.

By considering the new output, the relative of the system (17) is 3 at any point x_0 ((relative degree of the system is not well defined). The system (17) in normal form with respect to output z_1

$$\begin{aligned} \dot{z_1} &= z_2 \\ \dot{z_2} &= z_3 \\ \dot{z_3} &= a(z) + b(z)u \\ \dot{\eta} &= -\eta + z_3, \end{aligned} \qquad (19)$$

with $a(z) = x_4 - x_1^2x_3 - 3x_1x_3 - x_2x_3 - x_1x_4$, $b(z) = 1 + x_1$. Thus the system (17) is the minimum phase with respect to the new output.

Then according to (9), the modified steepest descents control with respect to z_1 is

$$\dot{u} = -2(1 + x_1)a_3\Big(a_0(z_1 - z_{1d}) + a_1(\dot{z_1} - \dot{z_{1d}}) + a_2(\ddot{z_1} - \ddot{z_{1d}}) + a_3\left(z_1^{(3)} - z_{1d}^{(3)}\right)\Big) + v, \qquad (20)$$

where v as in equation (10).

Let $y_d(t) = x_{1d}(t) = 0.5sin(t)$. Next, we choose $z_{1d}(t)$ such that if $z_1(t)$ tracks $z_{1d}(t)$, then $y(t)$ tracks the desired output $y_d(t)$. By replacing x_1 with $x_{1d}(t) = 0.5sin(t)$, then $x_{2d} = 0.5cos(t)$. By replacing x_2 with $x_{2d}(t)$, then $x_{3d} = -\frac{0.5sin(t)}{1+0.5sin(t)}$. By replacing x_3 with $x_{3d}(t)$, we have a differential equation $\dot{x_4} - x_4 = \frac{-0.5cos(t)}{(1+0.5sin(t))^2} + \frac{0.25sin^2(t)}{1+0.5sin(t)}$.

Thus $x_{4d} = 1/2(-0.5cos(t) - 0.5sin(t) + \frac{sin(t)}{1+0.5sin(t)})$. Now, $z_{1d} = 0.5cos(t)$.

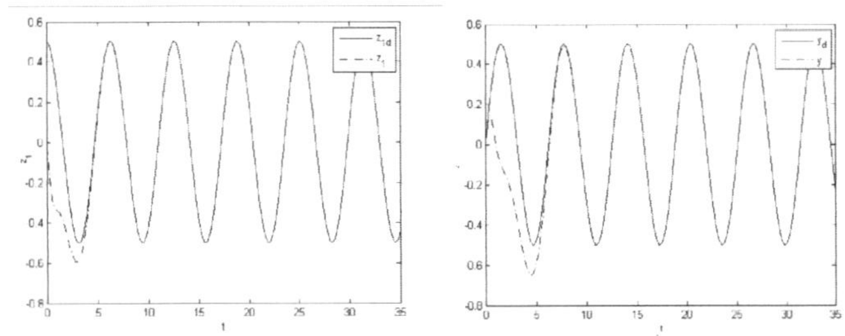

Figure 2. Left : Output Tracking, z_1 *to* z_{1d}; Right: Output Tracking (original system) y *to* y_d

Simulation results for the modified steepest descent control (20) are shown in Figure 2 for constants $a_0 = 15$, $a_1 = 23$, $a_2 = 9$, $a_3 = 1$. Initial value $x_1(0) = 0$, $x_2(0) = 0.5$, $x_3(0) = 0$, $x_4(0) = -0.5$, $u(0) = 0.2$:

3.3. Relative degree $r = n - 1$ with uncertainty

Consider the following SISO affine nonlinear control system

$$\begin{aligned} \dot{x_1} &= x_2 - x_1^3 \\ \dot{x_2} &= x_3 - u + 2x_1^3 \\ \dot{x_3} &= \theta\sin(x_1) + u - 2x_1^3 \end{aligned} \quad (21)$$

$$y = x_1. \quad (22)$$

The zero dynamic system (21)-(22) is $\eta = \dot{\eta}$. Thus the system (21)-(22 is the non-minimum phase.

Now redefining output : $z_1 = \alpha x_1 + x_2 + x_3$, with $0 < \alpha < 1$. Then we have the zero dynamic system (21)-(22) with respect to the output z_1 is

$$\dot{\eta} = \eta - \left(\frac{-\eta}{\alpha - 1}\right) - \left(\frac{-\eta}{\alpha - 1}\right)^3 + \theta\sin\left(\frac{-\eta}{\alpha - 1}\right),$$

and

$$\begin{aligned} \eta\dot{\eta} &= \eta^2 + \frac{\eta^2}{\alpha - 1} + \frac{\eta^4}{(\alpha - 1)^3} + \eta\theta\sin\left(\frac{-\eta}{\alpha - 1}\right) \\ &\leq \eta^2 + \frac{\eta^2}{\alpha - 1} + \frac{\eta^4}{(\alpha - 1)^3} + |\eta||\theta|\left|\frac{-\eta}{\alpha - 1}\right| \\ &= \frac{\eta^2(|\theta| - \alpha)}{|\alpha - 1|} + \frac{\eta^4}{\alpha - 1}. \end{aligned} \quad (23)$$

If $|\theta| \leq \alpha$, then $\eta\dot{\eta} < 0$. Thus the system (21) with respect to the output z_1 is minimum phase. Let $y_d(t) = \pi/2$. By replacing x_1 with $x_{1d} = y_d = \pi/2$, then $x_{2d} = (\pi/2)^3$. By replacing x_2 with x_{2d}, we have a differential equation $\dot{x_3} - x_3 = \theta$. Thus $x_{3d} = -\theta$. Now, $z_{1d} = \alpha x_{1d} + x_{2d} + x_{3d} = \alpha(\pi/2) + (\pi/2)^3 - \theta$.

The modified steepest descent control with respect to the output z_1 is

$$\dot{u} = -\frac{\partial F}{\partial u} = -2a_2(a_0(z_1 - z_{1d}) + a_1(\dot{z_1} - \dot{z_{1d}}) + a_2(\ddot{z_1} - \ddot{z_{1d}}))(1 - \alpha) + v, \quad (24)$$

where v as in equation (10).

Simulation results are shown in Figure 3 for constants $a_0 = 12$, $a_1 = 14$, $a_2 = 6$, $\alpha = 0.75$. Initial value $x_1(0) = 0,5$, $x_2(0) = 1$, $x_3(0) = 0$, $u(0) = 0$, $\theta(t) = 0.6$.

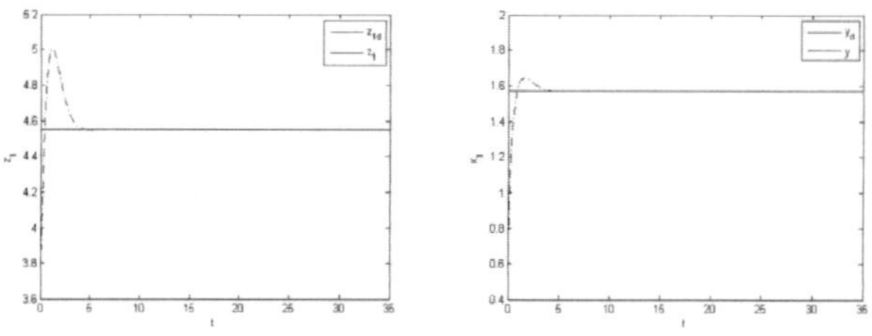

Figure 3. Left : Output Tracking, x_3 *to* x_{3d}; Right: Output Tracking (original system) y *to* y_d

4. Conclusion

In this paper, we have applied the input-output linearization and modified gradient descent control for 3 examples of the nonlinear nonminimum phase systems with or without uncertainty. Before applying both of method, the output of the system must be redefined as a linear combination of the state variables systems, such that the system becomes minimum phase with respect to a new output. Furthermore, the new desired output will be set based on the desired output of the original system. Simulation results are shown that the output of the systems tracks the output desired of the systems. From these results, it is possible to apply both methods to any relative degree of the nonlinear control system.

Acknowledgment

This research is supported by the Indonesian Directorate General of Higher Education (DIKTI)

Refrences

[1] A. Isidori. *Nonlinear Control Systems: An Introduction.* Springer, Berlin, Heidelberg, 1989.

[2] J. Naiborhu. *Output Tracking of Nonlinear Non-minimum Phase Systems by Gradient Descent Control.* Proceeding of the IASTED International Conference Identification, Control, and Applications (ICA-2009),August 17-19, 2009, Honolulu Hawaii, USA, 110-115

[3] Riccardo Marino. and Patrizio Tomei. A class of Globally Output Feedback Stabilizable Nonlinear Nonminimum Phase Systems. *IEEE Transactions on Automatic Control* **50** (12) (2005) 2097-2101.

[4] Z.Y. Zili Li. and Zengqiang Chen. The stability analysis and control of nonminimum phase nonlinear systems. *International journal of Nonlinear Science* **3** (2) (2007) 103-110.

[5] F. C. N. Wang, W. Xu. Adaptive global output feedback stabilisation of some non-minimum phase nonlinear uncertain systems. *IET control Theory Appl* **2** (2) (2008) 117-125.

[6] J.Naiborhu, Firman and K. Mu'tamar. Particle Swarm Optimization In The Exact Linearization Technic For Output Tracking of Non-Minimum Phase Nonlinear Systems. *Applied Mathematical Science* **7** (1) (2013) 5427-5442.

[7] Firman, J.Naiborhu and Roberd Saragih. Modification of a Steepest Descent Control for Output Tracking of Some Class Non-minimum Phase Nonlinear Systems. *Applied Mathematics and Computation* **269** (2015) 497-506.

[8] Dimitar Ho and J.Karl Hedrick. *Control of Nonlinear Non-Minimum Phase Systems with Input-Output Linearization,* Proceeding of American Control Conference, July 1-3, 2015. Chicago, IL, USA, 4016-4023

IOP Conf. Series: Journal of Physics: Conf. Series **1016** (2018) 012010 doi:10.1088/1742-6596/1016/1/012010

Parametric synthesis of a robust controller on a base of mathematical programming method

I.V. Khozhaev S.A. Gayvoronskiy and T.A. Ezangina

Department of Automatics and Computer Systems, National Research Tomsk Polytechnic University, Lenin Avenue 30, Tomsk, Russia

Email: eza-tanya@yandex.ru

Abstract. Considered paper is dedicated to deriving sufficient conditions, linking root indices of robust control quality with coefficients of interval characteristic polynomial, on the base of mathematical programming method. On the base of these conditions, a method of PI- and PID-controllers, providing aperiodic transient process with acceptable stability degree and, subsequently, acceptable setting time, synthesis was developed. The method was applied to a problem of synthesizing a controller for a depth control system of an unmanned underwater vehicle.

1. Introduction

Real control objects are commonly described with a set of parameters, which vary on the base of unknown laws or their accurate measuring is unavailable. Control systems, which manipulate such objects, are considered as interval control systems (ICS) or systems with interval parameters [66]. ICS can be described via its interval characteristic polynomial (ICP), which coefficients include interval parameters of a control object. ICP are widely used to perform ICS controller synthesis on a base of maximal operating speed criteria. It is known, that in order to maximize operating speed of the system, it is necessary to maximize robust stability degree. The problem of robust stability degree maximization was considered in papers [1-5], [8, 9].

There are many control systems, which require an aperiodic transient process. This makes a problem of providing simultaneously a maximal operating speed and an aperiodic transient process in ICS highly relevant. Aperiodic transient process in an ICS can be provided by allocating one migrating real pole closely to imaginary axis and allocating all other poles away of imaginary axis. This approach is based on a pole domination principle and is described in [10-13].

However, these papers describe a synthesis procedure for systems with interval ICP coefficients uncertainty. This makes relevant a problem of parametric synthesis of controllers for systems with affine uncertainty of ICP coefficients, in which primary interval parameters are included. Considering this, the main aim of the research is to develop a method of parametric synthesis of a robust controller, which would provide a maximal aperiodic stability degree for an ICS with affine uncertainty of ICP coefficients.

2. Algorithm of synthesizing a robust controller for an ICS

Let us consider an ICS with unity feedback, which includes a control object $W_{co}(s) = \frac{B(s)}{A(s)} = \frac{\sum_{h=0}^{c} b_h([\vec{T}])s^h}{\sum_{q=0}^{w} a_q([\vec{T}])s^q}$, here $[\vec{T}]$ – vector of interval, s – Laplace operator. Also, the system includes a controller $W_c(s,\vec{k}) = \frac{F(s,\vec{k})}{s}$, here $\vec{k}$ – vector of controller parameters. Considering this, an ICP of the system can be written as follows:

$$D\left(s,\overrightarrow{[T]},\vec{k}\right) = B(s,\overrightarrow{[T]})F(s,\vec{k}) + sA(s,\overrightarrow{[T]}) = \sum_{i=0}^{n} d_i(\overrightarrow{[T]},\vec{k})s^i \,. \tag{1}$$

It is proposed to solve the synthesis problem with the help of mathematical programming theory. Such approach is described in [5-7] and allows to maximize a stability degree α of a time-invariant system by synthesizing a vector of controller parameters $\vec{k}$ on the base of characteristic polynomial. Interval expansion of a mathematical programming method is proposed to be based on examining vertices characteristic polynomials (VCP), which is ICP in vertices of its parametric polytope:

$$D^q(s) = \sum_{i=1}^{m} T_i^q \cdot A_i(s) + B(s)\,, \tag{2}$$

here q – number of vertex.
Let us substitute s in (2) with $s = \alpha + j\beta$ and rewrite it as a complex number:

$$D^q\left(\vec{k},\alpha,\beta\right) = \operatorname{Re} D^q\left(\vec{k},\alpha,\beta\right) + \operatorname{Im} D^q\left(\vec{k},\alpha,\beta\right) \tag{3}$$

In order to find the maximal stability degree and parameters of a controller, a system of equations can be derived on the base of (3):

$$\begin{cases} \operatorname{Re} D^q\left(\vec{k},\alpha,\beta\right) = 0; \\ \operatorname{Im} D^q\left(\vec{k},\alpha,\beta\right) = 0; \\ \partial \operatorname{Re} D^q\left(\vec{k},\alpha,\beta\right) / \partial\alpha = 0; \\ \partial \operatorname{Im} D^q\left(\vec{k},\alpha,\beta\right) / \partial\alpha = 0; \\ \ldots \\ \partial^c \operatorname{Re} D^q\left(\vec{k},\alpha,\beta\right) / \partial\alpha^c = 0; \\ \partial^c \operatorname{Im} D^q\left(\vec{k},\alpha,\beta\right) / \partial\alpha^c = 0. \end{cases}$$

Let us notice, that in order to provide an aperiodic transient process, one of system poles, which is the closest to imaginary axis, should have zero imaginary part $\beta = 0$. Considering this, to provide an aperiodic transient process in considered ICS, a following system of equations must be solved:

$$\begin{cases} \operatorname{Re} D^q\left(\vec{k},\alpha\right) = 0; \\ \partial \operatorname{Re} D^q\left(\vec{k},\alpha\right) / \partial\alpha = 0; \\ \ldots \\ \partial^c \operatorname{Re} D^q\left(\vec{k},\alpha\right) / \partial\alpha^c = 0. \end{cases} \tag{4}$$

Considered research resulted in two methods of robust stability degree maximization for systems with different number of interval parameters, determining a number of parametric polytope P_T vertices. If number of interval parameters is less or equal to 3, than the synthesis algorithm can be formulated as follows:

1. Define ICP of the system in a form (1)
2. On the base of ICP, with the help of $s = \alpha + j\beta$ substitution, derive a VCP in a form (2)
3. Rewrite VCP (2) as (3)
4. Solve systems of nonlinear equations (4) in each vertex of parametric polytope P_T and determine values of aperiodic stability degree α and according parameters of controller $\vec{k}$.
5. Choose maximal aperiodic stability degree $\alpha^* = \max \alpha_i$ and according controller parameters $\vec{k}$ from the set of ($\alpha_i, \vec{k}$) in each vertex of P_T.

If the number of interval parameters exceeds 3, than the method of synthesis is based on combining synthesis procedures for different types of ICP coefficients uncertainty. It includes following steps:

1. With the help of coefficient method [9] find parameters of controller $\vec{k}$, providing a quasimaximal stability degree $\eta_{\max}$.
2. Plot an boundary vertex route of parametric polytope [14], considering previously found parameters of the controller.
3. Find projections of a vertex route on a complex plane and determine a vertex, which projection is the closest to the imaginary axis.
4. Calculate values of maximal stability degree α^* and according parameters of a controller $\vec{k}$ by solving the system of nonlinear equations in a vertex, found in the step 5.
5. If the closest to imaginary axis projection is a projection of an inner point of the parametric polytope edge and not a vertex, than it is proposed to substitute coordinates of this point in an ICP in step 3 instead of vertex coordinates a calculate a value of interval parameter T_i, $i \in \overline{1,m}$.

3. Mathematical Model of the Submerging Velocity Control Loop

Considered AUV is shown in the figure 1.

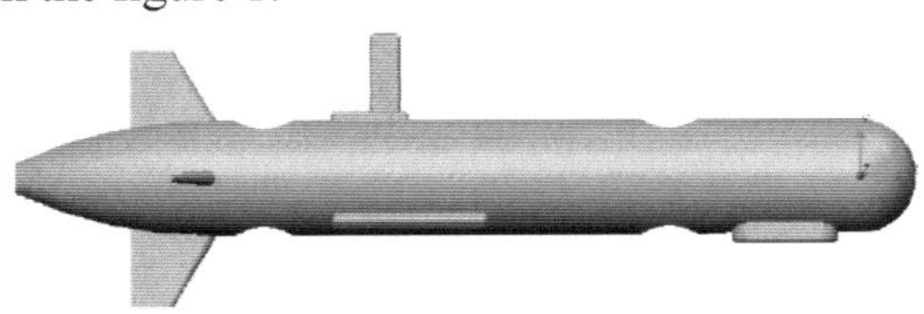

Figure 1. Hull of the considered AUV

Submerging control is performed with the help of two vertical thrusters. The structure of depth control channel is shown in the figure 2.

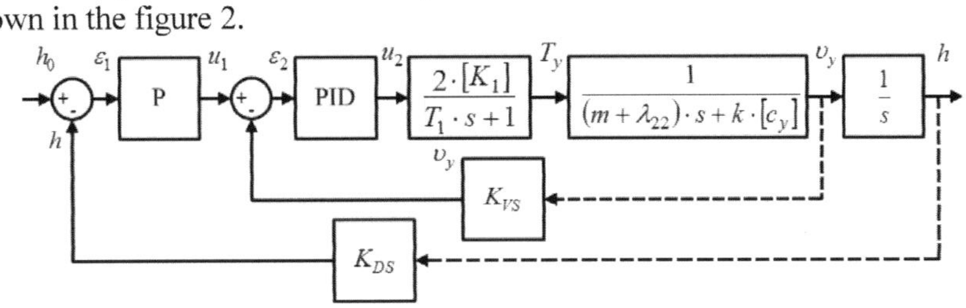

Figure 2. Structure of the submerging control channel

In the figure 2 following designations are accepted: h_0 – depth setpoint; h – actual value of depth; ε_1 – a difference between a setpoint and an actual value of the depth; P – P-controller of the outer control loop; u_1 – output of the P-controller; ε_2 – a difference between a P-controller output signal and a submerging velocity signal υ_y; PID – PID-controller of an inner control loop; K_1, T_1 – transfer coefficient and time constant of the thruster; T_y – thrust of vertical steering thrusters; m – AUV mass; λ_{22} – additional mass of water; c_y – hydrodynamic lift force coefficient; k – interval linearization coefficient; υ_y – submerging velocity; K_{VS} – transfer coefficient of the velocity sensor; K_{DS} – transfer coefficient of the depth sensor.

The problem of synthesizing a P-controller of the outer control loop will not be considered in the paper. Let us now consider the mathematical model of the inner control loop of the submerging control channel, which controls submerging velocity.

Considering previous designations, transfer function of the submerging velocity control loop can be written as follows:

$$W(s) = \frac{2 \cdot K_1 \cdot \left(K_D \cdot s^2 + K_P \cdot s + K_I\right)}{a_3 \cdot s^3 + a_2 \cdot s^2 + a_1 \cdot s + a_0}$$

$$a_0 = 2 \cdot K_1 \cdot K_I \cdot K_{VS};$$

$$a_1 = k \cdot c_y + 2 \cdot K_1 \cdot K_P \cdot K_{VS};$$

$$a_2 = m + \lambda_{22} + k \cdot T_1 \cdot c_y + 2 \cdot K_1 \cdot K_D \cdot K_{VS};$$

$$a_3 = G_1 \cdot \left(m + \lambda_{22}\right).$$

here K_P, K_I, K_D – proportional, integral and differential coefficients of the PID-controller.

The model considers an interval uncertainty of the AUV hydrodynamic parameters and interval uncertainty of the transfer coefficient of its thrusters. Тогда интервальный характеристический полином имеет вид

$$D(\mathrm{s}) = d_3 \cdot s^3 + [d_2] \cdot s^2 + [d_1] \cdot s + [d_0], \tag{5}$$

here $d_3 = a_3$, $d_2 = a_2$, $d_1 = a_1$, $d_0 = a_0$

4. PID-controller parametric synthesis

Let us perform a PID-controller synthesis via developed method. To do this, let rewrite (5) in an affine form:

$$D(\mathrm{s}) = [T_1] \cdot A_1(s) + [T_2] \cdot A_2(s) + B(s), \tag{6}$$

here $[T_1] = [c_y]$, $[T_2] = [K_1]$, $A_1 = k \cdot T_1 \cdot s^2 + k \cdot s$,

$A_2 = 2 \cdot (K_D \cdot K_{VS} \cdot s^2 + K_P \cdot K_{VS} \cdot s + K_I \cdot K_{VS})$, $\mathrm{B(s)} = G_1 \cdot \left(m + \lambda_{22}\right) \cdot s^3 + \left(m + \lambda_{22}\right) \cdot s^2$.

Expression (6) shows us, that parametric polytope P_T includes two interval parameters and, subsequently, 4 vertices. For each vertex of P_T let us derive a system (4). From set of solutions of all these systems, let us choose a maximal aperiodic stability degree $\alpha^* = 3.5$ and according parameters $\vec{k} = (11.367\ 13.758\ 0.118)$ of a controller.

5. Conclusion

Considered research resulted in a method of synthesizing a PI- or PID-controllers, providing aperiodic transient process in a control system, despite its interval parametric uncertainty. The method is applicable for synthesizing controllers of ICS with affine uncertainty of ICP coefficients and various numbers of interval parameters.

The method was tested on a problem of synthesizing a controller for a depth control system of a UUV.

IOP Conf. Series: Journal of Physics: Conf. Series **1016** (2018) 012010 doi:10.1088/1742-6596/1016/1/012010

References

[1] Chyi H Hwang J H Leu J F 2004 Tuning PID controllers for time-delay processes with maximizing the degree of stability Proc. of the 5th Asian Control Conference (*Australia*) vol 1 (Melbourne:Victoria) pp 466-471.

[2] Fiodorov I Synthesis algorithms of controllers for automatic control systems with maximum stability *Annals of the University of Craiova Electrical Engineering series* 2013 **37** pp. 139-143.

[3] Volkov A N, Zagashvili Y V The method of synthesis of automatic control systems with maximum stability under constraints *Izvestiya RAS, Ser. Computer and Systems* 1997 *3* pp. 12-19.

[4] Kim D P Synthesis regulator maximum degree of stability *Drive Technology* 2003 **4** pp.52-57

[5] Tatarinov A V Tsirlin A.M Mathematical programming problems containing complex variables, and the maximum degree of stability of linear dynamical systems *Bulletin RAS. Ser. Computer and Systems* 1995 1 pp. 28-33.

[6] Kochegurova, E.A., Gorokhova, E.S. Current estimation of the derivative of a nonstationary process based on a recurrent smoothing spline *Optoelectronics, Instrumentation and Data Processing*, 2016 **52 (3)**, pp. 280-285.

[7] Kochegurova, E., Gorokhova, E. Current derivative estimation of non-stationary processes based on metrical information Lecture Notes in Computer Science (including subseries Lecture Notes in Artificial Intelligence and Lecture Notes in Bioinformatics), **9330** LNCS 2015, pp. 512-519.

[8] Nesenchuk A.A. 2002 Parametric synthesis of qualitative robust control systems using root locus fields Proceedings of the 15th Triennial World Congress of The International Federation of Automatic Control (IFAC) Barcelona, Spain 21–26 July 2002 pp. 387-387.

[9] Ezangina T A, Gayvoronskiy S A The Control of the Descent Submersible in the Confused Sea Environment *Applied Mechanics and Materials* 2014 **536-537** pp. 1200-1206.

[10] Fuller A T Conditions for aperiodicity in linear systems *Brit. J. Appl. Phys* 1955. **6** pp 195-198.

[11] Meerov M V On aperiodicity robustness *International Journal of Control* 1998 **70(2)** pp 193 - 201.

[12] Wang Y Schinkel M., Hunt K J PID and PID-like controller design by pole assignment within D-stable regions *Asian Journal of Control* 2002 **4(4)** pp 423-432.

[13] Yakovleva, E.M., Faerman, V.A. Features of control systems analysis with discrete control devices using mathematical packages *IOP Conference Series: Materials Science and Engineering* 2017 **177(1)** pp. 833-838.

[14] Vadutov O S, Gayvoronskiy S. A Application of edge routing to stability analysis of interval polynomials *Journal of Computer and Systems Sciences International* 2003 **42(6)** pp. 833-838.

Acknowledgments

The reported study is supported by the Ministry of Education and Science of Russian Federation (project #2.3649.2017/PCh).

IOP Conf. Series: Journal of Physics: Conf. Series **1016** (2018) 012011 doi:10.1088/1742-6596/1016/1/012011

An improved measurement system for FOG pure lag time with no changing of FOG work status

X Chen, J H Yang*, Y L Zhou and X W Shu

College of Optical Science and Engineering, Zhejiang University, Hangzhou 310027, China
State Key Laboratory of Modern Opitical Instrumentation (Zhejiang University), Hangzhou, Zhejiang, 310027, China

*Email: xxianhua@zju.edu.cn

Abstract. The minimum pure lag time is an important factor for characterizing the dynamic performance of fiber optical gyroscope. It is defined as the time duration from the reception of velocity-shock signal to the output of corresponding fiber-optic gyroscope data. Many engineering projects have required for this index specifically, so the measurement of the minimum pure lag time is highly demanded. In typically measurement system, the work status of tested FOG has to be changed. In this work, a FOG pure lag time measurement system without changing the work status of the FOG has been demonstrated. During the operation of this test system, the impact structure generated a shock towards the FOG, and the pure lag time was measured through data processing analysis. The design scheme and test principle have been researched and analyzed in detail. And a prototype has been developed and used for experiment successfully. This measurement system can realize a measurement accuracy of better than $\pm 3\mu s$ and a system resolution of $108.6ns$.

1. Introduction

Fiber-optic gyroscope (FOG) is an angular velocity sensor based on the Sagnac effect. It has shown great advantages such as small volume, light weight, wide range of accuracy, and no need for moving parts[1], and has been extensively applied in defense applications such as aviation, spaceflight, navigation, and weapons[2,3].

For the development of high speed FOG inertial navigation system[4], the pure lag time is a key index. The FOG pure lag time is defined as: the time duration from the reception of velocity-shock signal to the output of corresponding FOG data.

The minimum pure lag time directly reflects the instantaneity of FOG as well as its feedback ability for large dynamic angular velocity signal. As more and more attentions has been paid to rapid response weapons research projects[5-7], the index of FOG pure lag time has been required specially in such projects. Since pure lag time is related with FOG's work status (such as temperature, communication frame rate, electromagnetic environment etc.), keeping the FOG at the normal work status is very important to get an accurate pure lag time[8,9].

In this paper, a convincing measurement system has been realized for FOG's pure lag time measurement. This measurement system realized higher measuring accuracy in a convincing method with sampling rate equal to normal frame rate of FOG. The minimum pure lag time of a typical FOG product has been measured and error of the system has been analyzed.

2. Design of the measurement system

Before demonstrating pure lag time measurement system, it is necessary to introduce FOG data test system. FOG data test system is a proven system for quality test of FOG products, which collects data from tested FOG working normally[10].

Since in our system, FOG' s work status should remain as usual during the measurement, it becomes a solving idea that convincing pure lag measurement system can be developed based on the FOG data test system.

2.1. The scheme of pure lag time measurement system

As shown in figure 1(a), in a FOG data test system, the data from FOG is firstly preprocessed by data collection module, and then sent to upper computer for calculation. Figure 1(b) shows the data transmission scheme of pure lag time measurement system. Each time a frame of FOG data has been received, a time information would been measured, which reflects the time duration from the generation of velocity-shock signal. Then the frame of FOD data and the measured time information will be packaged together and sent to upper computer (as will discussed in Section 2.4) .

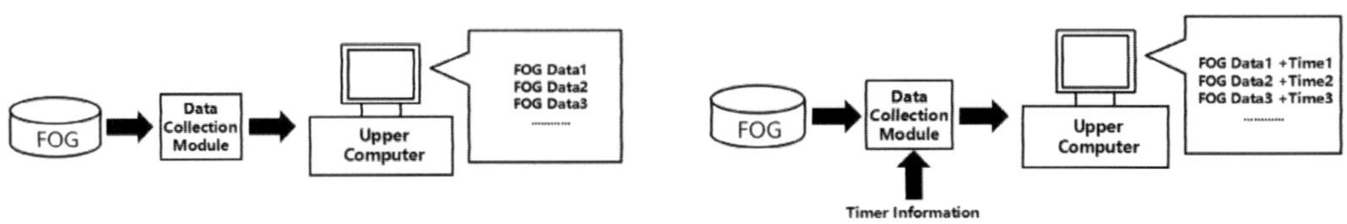

(a) (b)

Figure 1. Comparison of pure lag measurement system and data test system. (a) FOG data test system. (b) FOG pure lag time measurement system.

2.2. Principle of pure lag time measurement system

2.2.1 Establishing the time model of pure lag time measurement. Usually in the application of FOG, the FOG outputs data in fixed frame rate in their application, and the time interval for each frame of data is Δt_f .

Then the time model of pure lag time test for FOG can be established as shown in figure 2. The corresponding moments that data collection module receives FOG data $D_0, D_1, D_2 \dots D_n$ are t_0, t_1,

$t_2 \dots t_n$ respectively. The angular velocity impact happens at t_p, which is located between t_0 and t_1. Then we can get

$$\Delta t_f = t_1 - t_0 \tag{1}$$

$$\Delta t_a = t_p - t_0 \tag{2}$$

$$\Delta t_b = t_1 - t_p \tag{3}$$

where Δt_a denotes the time interval between t_p and the previous data reception moment t_0. And Δt_b denotes the time interval between t_p and the later data reception moment t_1.

Because of the existence of the time duration from the reception of velocity-shock signal to the output of corresponding FOG data(pure lag time), after the angular velocity impact happens, the previous FOG data (D_1 to D_{i-1}) remains unchanged until D_i is received, which appears more than twice as large as D_{i-1}. Then it can be judged that the angular velocity impact has been firstly reflected by the received data D_i. And The pure lag time, Δt_d, can be described as:

$$\Delta t_d = t_i - t_p = \Delta t_b + (i-1)\Delta t_f \tag{4}$$

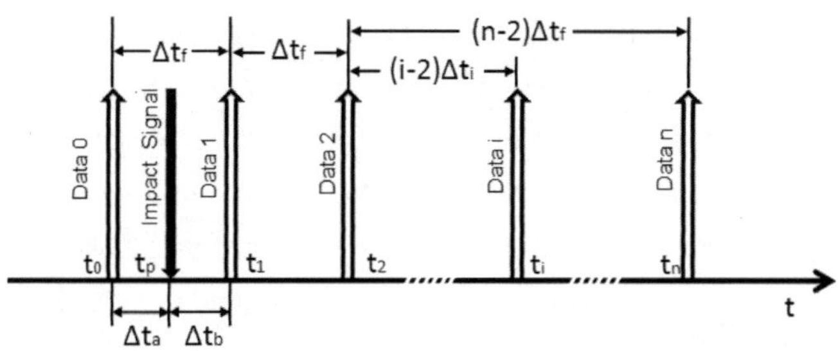

Figure 2. Principle model of FOG pure lag measurement.

2.2.2 Pure lag time measurement principle. For this system, during a pure lag time measurement, the position of t_p is random. The value range of Δt_b can be described as:

$$0 < \Delta t_b < \Delta t_f \tag{5}$$

There is a minimum pure lag time Δt_{dmin} for a certain work status of FOG. It is the minimum time duration from the reception of velocity-shock signal to the output of corresponding FOG data. Δt_{dmin} is the true value of minimum pure lag time and here $\Delta t_{dmin\text{-}result}$ is defined as the measurement result of Δt_{dmin}.

If the work status of tested FOG is determined, its minimum pure lag time Δt_{dmin} is also fixed. Here we determine a non-negative integer m as

$$\left\lfloor \frac{\Delta t_{d\min}}{\Delta t_f} \right\rfloor = m \tag{6}$$

There will be two different situations with different location of t_p between t_0 and t_1.

(1) In the case that

$$\Delta t_f > \Delta t_b > \Delta t_{d\min} - m\Delta t_f \tag{7}$$

Supposing that at the ti moment, the responding FOG data reflects the angular velocity impact, then the i can be determined as:

$$\mathrm{i} = m+1 \tag{8}$$

The pure lag time, Δt_d, can be described as:

$$\Delta t_d = \Delta t_b + m\Delta t_f \tag{9}$$

(2) While in another case that

$$\Delta t_{d\min} - m\Delta t_f > \Delta t_b > 0 \tag{10}$$

Supposing that at the ti moment, the responding FOG data reflects the angular velocity impact, then the i can be determined as:

$$i = m+2 \tag{11}$$

The measurement result of pure lag time, Δt_d, can be described as:

$$\Delta t_d = \Delta t_b + (m+1)\Delta t_f \tag{12}$$

2.2.3 Analysis of measurement result of pure lag time. Each time of measurement a pure lag time, Δt_d, can be gotten and the number of FOG data, i, which reflects the angular velocity impact. After enough times of measurement, a measured value of minimum pure lag time, $\Delta t_{dmin\text{-}result}$, can finally be gotten as described later in this article.

As mentioned in the previous section, a tested FOG with determined work status has a certain value of minimum pure lag time Δt_{dmin}. For the number i, there are two cases which correspond to i_1 and i_2 respectively (as determined by equation (8) and (11)). i_1 and i_2 meets the relationship below

$$i_2 - i_1 = 1 \tag{13}$$

And parameter m in equation(6) can be determined by equation(8) and equation(11)as follow:

$$m = i_1 - 1 \tag{14}$$

Or

$$m = i_2 - 2 \tag{15}$$

In the i_1 case, the value range of Δt_{dmin} can be determined by the measurement result Δt_d through equation (7) and equation (9):

$$\Delta t_{d\min} < \Delta t_d \tag{16}$$

In the i_2 case, the value range of Δt_{dmin} can be determined by the measurement result Δt_d through equation (10) and equation (12):

$$\Delta t_{d\min} > \Delta t_d - \Delta t_f \tag{17}$$

According to equation(16) and equation(17), repeating the measurement for some times, the minimum value of measurement results of Δt_d in the case of $i = i_1$ can be described as

$$\Delta t_{dx} = [\{\Delta t_d|_{i=i_1}\}]_{\min} \quad (18)$$

And the maximum value of measurement results of $(\Delta t_d - \Delta t_f)$ in the case of $i = i_2$ can be described as

$$\Delta t_{dy} = [\{(\Delta t_d - \Delta t_f)|_{i=i_2}\}]_{\max} \quad (19)$$

Then Δt_{dmin} can be limited in a certain range as

$$\Delta t_{dx} > \Delta t_{d\min} > \Delta t_{dy} \quad (20)$$

Then $\Delta t_{dmin\text{-}result}$ can be determined as

$$\Delta t_{d\min-result} = \frac{\Delta t_{dx} + \Delta t_{dy}}{2} \quad (21)$$

The uncertainty of the measurement result of minimum pure lag time Δt_{dmin} can be described as below

$$\delta_{\text{result}} = \frac{\Delta t_{dx} - \Delta t_{dy}}{2} \quad (22)$$

The measurement result of Δt_{dmin} can be denoted by the form of $\Delta t_{dmin\text{-}result} \pm \delta_{result}$.

2.3. The design of trigger module

Trigger module is constituted of two parts: impact structure and trigger circuit. The impact structure consists of sliding block and impacted body. The impacted body is fastened to a rotary table. Once the sliding block is released and shocks the impacted body, the impacted body will drive the rotary table[11]. In this way, the angular velocity impact can be delivered to the tested FOG. There is an electrode at the contact segment of sliding block and impacted body, so electric pulse can be generated as soon as the angular velocity impact is delivered to the tested FOG. The trigger module can guarantee that the angular velocity impact and the electric pulse occur at the same time, in order to realize the precise record of angular impact moment.

2.4. Design of data collection and procession module

F28M35 of TI has been chosen as processing core of data collection and procession module. Figure 3 shows the working process of data collection and procession module.

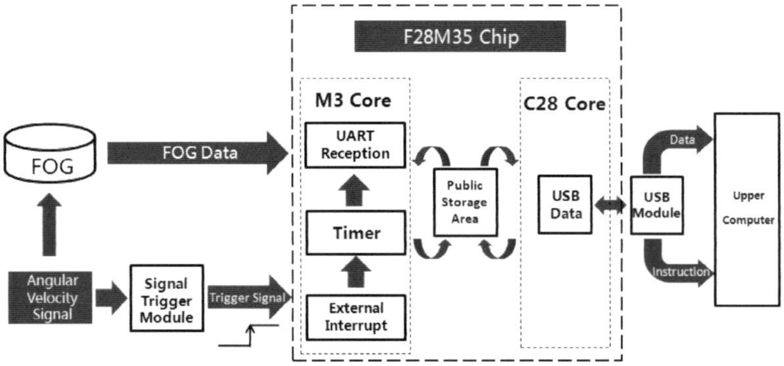

Figure 3. Working schematic diagram of data collection and processing module.

The packaged data received by upper computer consists of three parts: frame header, FOG data, and time information. The time information is determined by a 16-bits data, c. The most significant bit of time information is used as flag bit. For this measurement system, as long as c equals to 0xffff, the recorded time duration is determined. And the resolution of system is equivalent to timer resolution, which described as

$$\mu = \frac{\Delta t_{FS}}{2^{15}-1} \tag{23}$$

3. Realization of the measurement system

Pure lag time of FOG has been tested in the completed prototype of measurement system. The tested FOG works in broadcasting mode with a communication frame rate of 2KHz. The FOG has been tested for 10 times and the measurement result and analysis are explained below.

For this tested FOG, the time interval Δt_f between adjacent FOG data is $500\mu s$. The measurement results are shown in Table.1, which are arranged and numbered from small to large for convenience.

Table 1. measurement results of FOG pure lag time.

	Δt_d (μs)	i	Δt_d (μs)	Δt_d - Δt_f (μs)
1	244.83	i1	244.83	-
2	286.65	i1	286.65	-
3	354.81	i1	354.81	-
4	446.42	i1	446.42	-
5	486.91	i1	486.91	-
6	534.73	i2	-	34.73
7	581.28	i2	-	81.28
8	659.36	i2	-	159.36
9	702.59	i2	-	202.59
10	739.47	i2	-	239.47

As shown in the table, measurement results from group 1 to group 5 fit the case described by equation(7) and equation(8). According to equation (16), the upper limit of Δt_{dmin} can be get from corresponding Δt_d, and the result of these upper limits are arranged in the 1st column of Table.1.

Measurement results from group 6 to group 10 fit the case described by equation(10) and equation(11). According to equation (17), the lower limit of Δt_{dmin} can be get from their Δt_d, and the result of these lower limits are arranged in the 10th column of Table.1 .

3.1. Measurement results and measurement accuracy

From equation (18), Δt_{dx} can be gotten as

$$\Delta t_{dx} = [\{\Delta t_d|_{n=n_1}\}]_{\max} = 244.83\mu s \tag{24}$$

From equation (19), Δt_{dy} can be gotten as

$$\Delta t_{dy} = [\{\Delta t_d|_{n=n_2}\}]_{\max} = 239.47\mu s \tag{25}$$

According to equation (21), (24) and (25), the measurement result of Δt_{dmin} can be gotten as

$$\Delta t_{d\min-result} = \frac{\Delta t_{dx} + \Delta t_{dy}}{2} = 242.15\mu s \tag{26}$$

And the measurement uncertainty can be gotten from equation(22)(24)(25) as

$$\delta_{\text{result}} = \frac{\Delta t_{dx} - \Delta t_{dy}}{2} = 2.83\mu s \tag{27}$$

Therefore, the measurement result of minimum pure lag time can be described as

$$\Delta t_{d\min-result} \pm \delta_{\text{result}} = (242.15 \pm 2.83)\mu s \tag{28}$$

3.2. System resolution

The system resolution of a completed measurement system is certain. The recorded time duration Δt_{FS} is $3\,560\mu s$ once the c is set to 0xffff (as mentioned in Section 1.3). So the system resolution can be gotten as

$$\mu = \frac{\Delta t_{FS}}{2^{15}-1} = \frac{3\,560\mu s}{2^{15}-1} = 108.6ns \tag{29}$$

4. Conclusion

Keeping the FOG at the normal work status is very important to get an accurate test result of pure lag time. Based on the FOG data test system, a measurement system of FOG pure lag time has been designed and realized. This measurement system can realize accurate pure lag time measurement without changing the FOG's work status. In the construction of test system, trigger module was designed to generate angular-velocity impact, and test data was collected and analyzed by data collection and procession module. The prototype has been developed and the pure lag time of FOG product has been measured successfully. A measurement uncertainty of $\pm 2.83\mu s$ and system resolution of $108.6ns$ have been achieved. In the follow-up research, the system can be further improved by controlling the moment of angular velocity impact precisely by program, so that the high accuracy can be gotten easily and the testing efficiency can also be improved.

References

[1] Zhang G C 2008 The Principle and Technologies of Fiber-Optic Gyroscope (Beijing: National Defense Industry Press)

[2] Sanders G A, et al. 2016 Fiber optic gyroscope development at Honeywell Spie Comm. + Sci. Sens. Imaging 9852 985207.

[3] Minakuchi S, Sanada T, Takeda N, Mitani S, Mizutani T, Sasaki Y and Shinozaki K 2015 Thermal strain in lightweight composite fiber-Optic gyroscope for space application. J. Lightwave Technol. 33 2658-62.

[4] Jin J, Song N F and Zhang C X 2002 Design of high speed detection circuit for fiber-optic gyroscope J. Chinese Inertial Tech. 10 49-53.

[5] Zhou Y L 2015 Application of over modulation technique in fiber optic gyroscope north-seeking. J. Zhejiang Univ. (Engineering Science) 49 1817-1820.

[6] Wang Y Z, Ma L, Yu H, Gao H Y and Yuan Y J 2016 Int. Symp. Optoelectronic Tech. Appl. 10158 101580S

[7] Yoon Y G, Lee S M and Kim J H 2015 J. Inst. Control 21 168-172.

[8] Xue N, Liu R, Zhang C, Zhou H, Wang L, Shu X W and Chen X F 2011 A FOG output delay time measurement device and its measuring method. CN 201110090599.9

[9] Chen X, Zhou Y L, Yang J H and Shu X W 2015 A pure lag time measurement device and its measuring method. CN 201510612439.4

[10] Liu S, Chen M and Zhao L 2007 Design of fiber optical gyroscope test data collecting system Computer Measurement & Control 15 972-974.

[11] Wang Y Y, Zhou Y L, Yang J H, Chen X F and Liu C 2015 A programmable impact device for pure lag time test. CN201320612757.7

IOP Conf. Series: Journal of Physics: Conf. Series **1016** (2018) 012012 doi:10.1088/1742-6596/1016/1/012012

Distributed Nash Equilibrium Seeking for Generalized Convex Games with Shared Constraints

Chao Sun[1] **and Guoqiang Hu**[1]

[1] School of Electrical and Electronic Engineering, Nanyang Technological University, Singapore 639798.

E-mail: gqhu@ntu.edu.sg.

Abstract. In this paper, we deal with the problem of finding a Nash equilibrium for a generalized convex game. Each player is associated with a convex cost function and multiple shared constraints. Supposing that each player can exchange information with its neighbors via a connected undirected graph, the objective of this paper is to design a Nash equilibrium seeking law such that each agent minimizes its objective function in a distributed way. Consensus and singular perturbation theories are used to prove the stability of the system. A numerical example is given to show the effectiveness of the proposed algorithms.

1. Introduction

How to find a Nash equilibrium is an interesting and important problem for non-cooperative games [1]. In [2], an iterative steepest descent algorithm was proposed for numerical approximation of local Nash equilibria. In [3], an adaptive learning technique was used to iteratively compute the Nash equilibrium of a multi-player game. Extremum seeking based methods were proposed to search Nash equilibrium [4–6].

Compared with the conventional Nash equilibrium problem, the generalized Nash games assume that each player's feasible set can be constrained by the rival players' strategies [7–9]. A generalized convex game usually has continuous strategy spaces and the actions are coupled through both objective functions and constraints. The problem was first formally proposed by [10], followed by [11] studying economic equilibria. Since then, there have been many studies reported on the existence of a generalized Nash equilibrium, and algorithms to compute it. For instance, a method involving variational inequalities was presented in [12].

Most of the aforementioned literature require that each player can obtain the strategies of its opponents. In practice, especially in some multi-agent networks, the agent might have limited interaction (communication) with other agents. Recently, distributed computation algorithms have attracted much attention due to their reliability, security and low communication burden. For example, [13–16] addressed the distributed optimization problem using saddle point dynamics methods and [17] investigated a distributed time-varying optimization problem for quadratic objective functions. The recently proposed algorithm in [18] solved a distributed Nash equilibrium seeking problem for unconstrained non-cooperative games based on average consensus and singular perturbation theory. In [19], discrete-time adaptive algorithms were presented to solve Nash equilibrium seeking problems, the objectives and constraints are required to be neighbor-coupled.

In this paper, we present a continuous-time distributed algorithm to seek a Nash equilibrium for a generalized convex game with shared constraints. The main contributions of this paper are summarized as follows: 1) A distributed algorithm to find the normalized Nash equilibrium of a

IOP Conf. Series: Journal of Physics: Conf. Series **1016** (2018) 012012 doi:10.1088/1742-6596/1016/1/012012

generalized convex game with class-C^2 objective functions is presented. Compared with the previous work [18], constraints relying on other players' strategies are considered. Compared with [19], the objective functions and constraints can be coupled arbitrarily; 2) The convergence of the players actions to the normalized Nash equilibrium is analyzed, by using singular perturbation based techniques, it is proven that the proposed algorithms converge into a neighborhood of the Nash equilibrium and the error bound can be arbitrarily small by selecting the control parameters.

The rest of this paper is organized as follows: In Section 2, notations and background information on graph theory are given. In Section 3, the distributed Nash equilibrium seeking problem is formulated mathematically. In Section 4, a distributed algorithm is designed, and convergence analysis is given. Section 5 gives a numerical example to illustrate the effectiveness of the proposed algorithms. Finally, Section 6 concludes the paper.

2. Notations and Preliminaries

Throughout this paper, R ($R^{\geq 0}$), R^n and $R^{n\times n}$ denote the set of reals (nonnegative reals), n-dimensional column vectors, and $n \times n$ matrices, respectively. **1** and **0** are column vectors with appropriate dimension. The symbol $\left[a_i\right] \in R^N$ represents a column vector defined as $\left[a_i\right] = [a_1,\ldots,a_N]^T$. A function f: $\left[0,a\right) \rightarrow \left[0,\infty\right)$ is said to belong to class-K if it is continuous, zero at zero, and strictly increasing. It is said to belong to class-K_∞ if it belongs to class-K, $a=\infty$ and $\lim_{r\to\infty} f(r)=\infty$. A function $f(x)$ is said to belong to class- C^2 if the derivatives $f^{(1)}(x), f^{(2)}(x)$ exist and are continuous.

Let $\mathrm{G}=\left\{\mathrm{V},\mathrm{E}\right\}$ denote an undirected graph, where $\mathrm{V}=\left\{1,\ldots,N\right\}$ indicates the vertex set and $\mathrm{E}\subset \mathrm{V}\times\mathrm{V}$ indicates the edge set. $\mathrm{N}_i=\left\{j\in\mathrm{V} \mid (j,i)\in\mathrm{E}\right\}$ denotes the neighborhood set of vertex i. A path is referred by the sequence of its vertices. Path P between v_0 and v_k is the sequence $\left\{v_0,\ldots,v_k\right\}$ where $(v_{i-1},v_i)\in\mathrm{E}$ for $i=1,\ldots,k$ and the vertices are distinct. The number k is defined as the length of path P. Graph G is connected if for any two vertices, there is a path in G. A matrix $A=\left[a_{ij}\right]\in R^{N\times N}$ denotes the adjacency matrix of G, where $a_{ij}>0$ if and only if $(j,i)\in\mathrm{E}$ else $a_{ij}=0$. In this paper, we suppose $a_{ii}=0$. A matrix $L\,\square\, D-A\in R^{N\times N}$ is called the Laplacian matrix of G, where $D=\mathrm{diag}\{d_{ii}\}\in R^{N\times N}$ is a diagonal matrix with $d_{ii}=\sum_{j=1}^{N}a_{ij}$.

3. Problem Formulation

Consider the set of players $\mathrm{V}\,\square\,\{1,\cdots,N\}$ where the action of player i is denoted as $x_i\in R$. Each player i minimizes its objective function $f_i(x):R^N\rightarrow R$ where $x=\left[x_1,\cdots,x_N\right]^T\in R^N$ is the action vector of N players. Suppose that each player is subject to k shared constraints that depend on the players' actions, i.e., $g_j(x)\leq 0,\quad j\in\mathrm{K}\,\square\,\left\{1,\ldots,\kappa\right\}$. If agent j is not a neighbor of agent i, then player i has no access to player j's action directly. Otherwise, player i can get the information of player j via a connected undirected graph topology. Our objective is to design a distributed Nash equilibrium seeking law for the players such that their actions converge to a Nash equilibrium of the game.

Denote $x_{-i}=\left[x_1,\cdots,x_{i-1},x_{i+1},\cdots,x_N\right]^T\in R^{N-1}$. Then, for player $i\in\mathrm{V}$, its optimization problem can be described as

$$\min_{x_i}\ f_i(x_i,x_{-i}),\ \text{such that}\ g_j(x_i,x_{-i})\leq 0,\ j\in\mathrm{K}. \tag{1}$$

IOP Conf. Series: Journal of Physics: Conf. Series **1016** (2018) 012012 doi:10.1088/1742-6596/1016/1/012012

Throughout this paper, suppose that the following assumptions on the constraints always hold.

Assumption 1 *$g_j(x): R^N \to R$ is a class-C^2 function and is convex with respect to x for every $j \in \mathrm{K}$.*

Assumption 2 *The constraint set $\mathrm{C} = \{x \in R^N | \ g_j(x) \le 0, \ j \in \mathrm{K}\}$ is nonvoid and bounded. Furthermore, the Slater's conditions hold.*

4. Distributed Generalized Nash Equilibrium Seeking Law Design and Convergence Analysis

In this section, we aim to find a Nash equilibrium of a generalized convex game, where the objective functions satisfy the following assumption.

Assumption 3 *$f_i(x_i, x_{-i}): R^N \to R$ is a class- C^2 function and is convex with respect to x_i for every $i \in \mathrm{V}$.*

4.1. Existence and Uniqueness of the Normalized Nash Equilibrium

Under Assumptions 1-3, the players' optimization problems are convex and Nash equilibria exist [7]. Moreover, a point $x^* = \left[x_1^*, \cdots, x_N^*\right]^T \in R^N$ is a Nash equilibrium of the game if and only if the following KKT conditions hold for all $i \in \mathrm{V}$ and some $\lambda_{ij}^* \ge 0, \ j \in \mathrm{K}$:

$$\nabla_{x_i} f_i(x^*) + \sum\nolimits_{j \in \mathrm{K}} \lambda_{ij}^* \nabla_{x_i} g_j(x^*) = 0, \lambda_{ij}^* g_j(x^*) = 0, \ g_j(x^*) \le 0, \ j \in \mathrm{K}. \tag{2}$$

Usually, the problem in (1) has multiple Nash equilibria. In this paper, we do not assume the uniqueness of the Nash equilibrium. Instead, we consider a special kind of Nash equilibrium, namely the normalized equilibrium point [7], which is defined according to the KKT condition (2), such that for all $i \in \mathrm{V}$,

$$\nabla_{x_i} f_i(x^*) + \sum\nolimits_{j \in \mathrm{K}} \lambda_j^* \nabla_{x_i} g_j(x^*) = 0, \lambda_j^* g_j(x^*) = 0, \ g_j(x^*) \le 0, \ j \in \mathrm{K}, \tag{3}$$

where $\lambda_j^* \ge 0$ is the common Lagrange multiplier for each shared constraint $g_j(x)$.

Before presenting the algorithm, we show that under some assumptions, the normalized Nash equilibrium satisfying (3) exists and is unique. Firstly, we introduce the concept of monotonicity for a single-valued mapping $F: R^N \to R^N$.

Definition 1 (On monotonicity of a single-valued mapping) *[19] (1) A mapping $F: R^N \to R^N$ is said to be monotone on R^N if for all $x, \hat{x} \in R^N, x \ne \hat{x}, (F(x) - F(\hat{x}))^T (x - \hat{x}) \ge 0$.*

(2) A mapping $F: R^N \to R^N$ is said to be strictly monotone on R^N if for all $x, \hat{x} \in R^N, x \ne \hat{x}, (F(x) - F(\hat{x}))^T (x - \hat{x}) > 0$.

Assumption 4 *[7] The function $\nabla_x f(x) = [\nabla_{x_1} f_1(x), \ \cdots, \nabla_{x_N} f_N(x)]^T$ is strictly monotone on R^N.*

Lemma 1 *[7] Under Assumptions 1-4, the normalized Nash equilibrium x^* satisfying (3) exists and is unique.*

The following assumption will be used in the stability analysis.

Assumption 5 *Under Assumptions 1-4, the pair $(x^*, \lambda_1^*, \cdots, \lambda_\kappa^*)$ satisfying (3) is unique.*

4.2. Control Design and Stability Analysis

We now propose a distributed control law using the estimation of neighboring actions. Let $Y_i = [y_{i1}, \cdots, y_{iN}]^T$ be player i's estimation on all the players' actions, which is produced by the following consensus based protocol

$$y_{ii} = x_i, i \in \mathrm{V}, \dot{y}_{ij} = -w_{ij} \sum_{k=1}^{N} a_{ik}(y_{ij} - y_{kj}), j \in \mathrm{V} / \{i\}, \tag{4}$$

with w_{ij} being a positive constant.

Based on (4), the updating law for play i is designed as

$$\dot{x}_i = -\bar{k}_i(\nabla_{x_i} f_i(Y_i) + \sum\nolimits_{j\in K} \nabla_{x_i} \lambda_{ij} g_j(Y_i)),$$
$$\dot{\lambda}_{1j} = \bar{k}_{1j}\lambda_{1j} g_j(Y_1), \dot{\lambda}_{ij} = -\gamma_{ij}\sum_{k=1}^{N} a_{ik}(\lambda_{ij} - \lambda_{kj}),\ j \in K, i \neq 1, \tag{5}$$

where player 1 is selected to calculate the common multipliers λ_{1j} and consensus based control laws are used to broadcast them to all the other players. In (5), $\lambda_{1j}(0) > 0$, $\gamma_{ij} > 0$, and $\bar{k}_i = \varepsilon k_i$, where ε is a small positive constant and $k_i > 0$ is a positive constant.

The distributed Nash equilibrium seeking algorithm in (4) and (5) uses leader-following consensus to estimate unknown information and saddle point dynamics to achieve convergence. In the following analysis, we use singular perturbation theory to prove the stability of the system. First, an auxiliary system is designed. Then, based on the stability of this auxiliary system, we prove the convergence of the original system using Lyapunov based methods.

For agent i, define a subgraph $G_i = \{V_i, E_i\}$, where $V_i = V / \{i\}$ and $E_i \subset V_i \times V_i$ indicate the set of vertices and edges, respectively. Let L_i be the Laplacian matrix of G_i and B_i = diag $\{a_{1i}, \cdots, a_{(i-1)i}, a_{(i+1)i}, \cdots, a_{Ni}\} \in R^{(N-1)\times(N-1)}$ is the information exchange matrix of agent i. Then, we can rewrite (4) and (5) as:

$$\dot{x} = \bar{\mathbf{k}}(\nabla_x f(Y) + \sum\nolimits_{j\in K} \text{diag}\{\lambda_{1j}, \boldsymbol{\lambda}_j\}\nabla_x g_j(Y)), \dot{\lambda}_{1j} = \bar{k}_{1j}\lambda_{1j} g_j(Y_1),$$
$$\dot{\boldsymbol{\lambda}}_j = -\Gamma_j(L_1 + B_1)(\boldsymbol{\lambda}_j - \lambda_{1j}\mathbf{1}), j \in K, \dot{\bar{Y}}_i = -W_i(L_i + B_i)(\bar{Y}_i - x_i\mathbf{1}), i \in V, \tag{6}$$

where $Y = [Y_1^T, \cdots, Y_N^T]^T \in R^{N^2}$, $\nabla_x f(Y) = [\nabla_{x_1} f_1(Y_1), \cdots, \nabla_{x_N} f_N(Y_N)]^T$, $\nabla_x g(Y) = [\nabla_{x_1} g(Y_1), \cdots, \nabla_{x_N} g(Y_N)]^T \in R^N$, $\boldsymbol{\lambda}_j = [\lambda_{2j}, \cdots, \lambda_{Nj}]^T$, $\bar{Y}_i = [y_{1i}, \cdots, y_{(i-1)i}, y_{(i+1)i}, \cdots, y_{Ni}]^T \in R^{N-1}$, $\bar{\mathbf{k}}$ = diag $\{\bar{k}_1, \cdots, \bar{k}_N\} \in R^{N\times N}$, Γ_j = diag $\{\gamma_{2j}, \cdots, \gamma_{Nj}\}$, W_i = diag $\{w_{1i}, \cdots, w_{(i-1)i}, w_{(i+1)i}, \cdots, w_{Ni}\} \in R^{(N-1)\times(N-1)}$. Let $[\zeta_j]$ be a column vector defined as $[\zeta_1^T, \cdots, \zeta_\kappa^T]^T$ and $[\eta_i]$ be a column vector defined as $[\eta_1^T, \cdots, \eta_N^T]^T$. Define an error variable $\Delta(t) = [(x(t) - x^*)^T$, $[\lambda_{1j}(t) - \lambda_{1j}^*]^T$, $[\boldsymbol{\lambda}_j(t) - \lambda_{1j}^*\mathbf{1}]^T$, $[\bar{Y}_i(t) - x_i^*\mathbf{1}]^T]^T$. The following theorem presents the main result of this section.

Theorem 1 *Suppose that Assumptions 1-5 hold and let (4) and (5) be the updating law. Then, for each pair of positive constants* $(r, \tilde{n})$, *there exists a positive constant* $\varepsilon^*(r, \tilde{n})$ *such that for every* $0 < \varepsilon < \varepsilon^*(r, \tilde{n})$, *there exists a time T such that* $\|\Delta(t)\| \leq \tilde{n}$, $\forall t \geq T$ *for every* $\|\Delta(0)\| \leq r$.

Proof: Firstly, introduce the following auxiliary system

$$\dot{x} = -\mathbf{k}(\nabla_x f(x) + \sum\nolimits_{j\in K} \lambda_{1j}\nabla_x g_j(x)), \dot{\lambda}_{1j} = k_{1j}\lambda_{1j} g_j(x), j \in K, \tag{7}$$

where $\mathbf{k} = \text{diag}\{k_1, \cdots, k_N\} \in R^{N\times N}$.

For (7), define a Lyapunov candidate function as

$$V_0 = \frac{1}{2}(x - x^*)^T \mathbf{k}^{-1}(x - x^*) + \sum_{j\in K}\frac{1}{k_{1j}}(\lambda_{1j} - \lambda_{1j}^* - \lambda_{1j}^* \log(\lambda_{1j}) + \lambda_{1j}^* \log(\lambda_{1j}^*)),$$

where $[x^{*T}, \lambda_{11}^*, \cdots, \lambda_{1K}^*]^T$ is the point satisfying (3). Taking the derivative of V along (7), we have

$$\dot{V}_0 = (x - x^*)^T(-\nabla_x f(x) - \sum_{j\in K} \lambda_{1j} \nabla_x g_j(x)) + \sum_{j\in K} (\lambda_{1j} - \lambda_{1j}^*) g_j(x). \quad (8)$$

Based on (3), if $\lambda_{1j}^* > 0$, we have $(\lambda_{1j} - \lambda_{1j}^*) g_j(x^*) = 0$; otherwise $\lambda_{1j}^* = 0$, which implies that $(\lambda_{1j} - \lambda_{1j}^*) g_j(x^*) \le 0$. Under Assumption 4, we have $(x - x^*)^T(\nabla_x f(x) - \nabla_x f(x^*)) \ge 0$. Using (3), equation (8) can be rewritten as

$$\begin{aligned}
\dot{V}_0 &\le -(x - x^*)^T \nabla_x f(x^*) - \sum_{j\in K} (x - x^*)^T \lambda_{1j} \nabla_x g_j(x) + \sum_{j\in K} (\lambda_{1j} - \lambda_{1j}^*)(g_j(x) - g_j(x^*)) \\
&= -(x - x^*)^T(-\sum_{j\in K} \lambda_{1j}^* \nabla_x g_j(x^*)) - \sum_{j\in K} (x - x^*)^T \lambda_{1j} \nabla_x g_j(x) + \sum_{j\in K} (\lambda_{1j} - \lambda_{1j}^*)(g_j(x) - g_j(x^*)) \\
&\le \sum_{j\in K} \lambda_{1j}^*(g_j(x) - g_j(x^*)) + \sum_{j\in K} \lambda_{1j}(g_j(x^*) - g_j(x)) + \sum_{j\in K} (\lambda_{1j} - \lambda_{1j}^*)(g_j(x) - g_j(x^*)) \\
&= 0,
\end{aligned}$$

where we utilize the convexity of $g_j(x)$.

Based on Assumption 4, letting $\dot{V}_0 \equiv 0$ gives $x = x^*$. Then, the largest invariant set can be written as

$$\begin{aligned}
S = \{&(x, \lambda_{11}, \cdots, \lambda_{1K}) \in R^N \times R^{\ge 0} \times \cdots \times R^{\ge 0} : \\
&\mathbf{0} = \nabla_x f(x) + \sum_{j\in K} \lambda_{1j} \nabla_x g_j(x), x = x^*, \lambda_{1j} g_j(x^*) = 0, j \in K\}.
\end{aligned}$$

It can be seen that all the points in S satisfy the condition in (3). According to Lemma 1 and Assumption 5, such a point is unique. Define $z = [(x - x^*)^T, [\lambda_{1j} - \lambda_{1j}^*]^T]^T$. Thus, for each $z \ne \mathbf{0}$, there exists a positive definite function $\beta(z)$ such that

$$(x - x^*)^T(-\nabla_x f(x) - \sum_{j\in K} \lambda_{1j} \nabla_x g_j(x)) + \sum_{j\in K} (\lambda_{1j} - \lambda_{1j}^*) g_j(x) < -\beta(z). \quad (9)$$

For notational convenience, $\hat{Y}_i = \bar{Y}_i - x_i \mathbf{1}$ and $\hat{\boldsymbol{\lambda}}_j = \boldsymbol{\lambda}_j - \lambda_{1j}\mathbf{1}$. Define a Lyapunov candidate $V(z, [\hat{Y}_i], [\hat{\boldsymbol{\lambda}}_j]) = cV_0 + (1-c)V_1$, where $c > 0$ is a positive constant and $V_1 = \frac{1-c}{2}(\sum_{i=1}^{N} \hat{Y}_i^T \hat{Y}_i + \sum_{j\in K} \hat{\boldsymbol{\lambda}}_j^T \hat{\boldsymbol{\lambda}}_j)$. It can be obtained that there exist two class-K_∞ functions α_1 and α_2 such that $\alpha_1(z, [\hat{Y}_i], [\hat{\boldsymbol{\lambda}}_j]) \le V(z, [\hat{Y}_i], [\hat{\boldsymbol{\lambda}}_j]) \le \alpha_2(z, [\hat{Y}_i], [\hat{\boldsymbol{\lambda}}_j])$.

Thus, taking the derivative of V along (6) gives

$$\begin{aligned}
\dot{V} &= -c\varepsilon(x - x^*)^T(\nabla_x f(Y) - \nabla_x f(x)) - c\varepsilon(x - x^*)^T \nabla_x f(x) - c\varepsilon(x - x^*)^T \\
&\times (\sum_{j\in K} \mathrm{diag}\{\lambda_{1j}, \boldsymbol{\lambda}_j\} \nabla_x g_j(Y) - \sum_{j\in K} \lambda_{1j} \nabla_x g_j(Y)) - c\varepsilon(x - x^*)^T \\
&\times (\sum_{j\in K} \lambda_{1j} \nabla_x g_j(Y) - \sum_{j\in K} \lambda_{1j} \nabla_x g_j(x)) - c\varepsilon(x - x^*)^T \sum_{j\in K} \lambda_{1j} \nabla_x g_j(x) \\
&- c\varepsilon \sum_{j\in K} (\lambda_{1j} - \lambda_{1j}^*)(g_j(Y) - g_j(x)) - c\varepsilon \sum_{j\in K} (\lambda_{1j} - \lambda_{1j}^*) g_j(x) \\
&- (1-c)\sum_{i=1}^{N} \hat{Y}_i^T W_i (L_i + B_i)\hat{Y}_i - (1-c)\sum_{i=1}^{N} \hat{Y}_i^T \frac{dx_i}{dt}\mathbf{1} - (1-c)\sum_{j\in K} \hat{\boldsymbol{\lambda}}_j^T \Gamma_j \\
&\times (L_1 + B_1)\hat{\boldsymbol{\lambda}}_j - (1-c)\sum_{j\in K} \hat{\boldsymbol{\lambda}}_j^T \frac{d\lambda_{1j}}{dt}\mathbf{1}.
\end{aligned}$$

Furthermore, according to Assumptions 1 and 3, for every Δ belonging to a compact set,

$$\dot V \le -c\varepsilon\beta(z) + cl_1\varepsilon\|x-x^*\|(\sum_{i=1}^{N}\|\hat Y_i\|^2)^{\frac{1}{2}} + c\varepsilon\|x-x^*\|\sum_{j\in K}(\|\hat{\boldsymbol\lambda}_j\|\|\nabla_x g_j(Y)\|)$$
$$+c\varepsilon l_2\kappa\|x-x^*\|(\sum_{i=1}^{N}\|\hat Y_i\|^2)^{\frac{1}{2}}\max\{\lambda_{1j}\} + c\varepsilon l_3\sum_{j\in K}\|\lambda_{1j}-\lambda_{1j}^*\|(\sum_{i=1}^{N}\|\hat Y_i\|^2)^{\frac{1}{2}}$$
$$-(1-c)l_4\sum_{i=1}^{N}\|\hat Y_i\|^2 - (1-c)\sum_{i=1}^{N}\hat Y_i^T\frac{dx_i}{dt}\mathbf{1} - (1-c)l_5\sum_{j\in K}\|\hat{\boldsymbol\lambda}_j\|^2$$
$$-(1-c)\sum_{j\in K}\hat{\boldsymbol\lambda}_j^T\frac{d\lambda_{1j}}{dt}\mathbf{1},$$

where $l_1,\cdots,l_5$ are some positive constants.

Let $\delta_1,\cdots,\delta_8$ be positive constants that can be arbitrarily chosen. Using Young's Inequality gives

$$\dot V \le -c\varepsilon\beta(z) - [(1-c)l_4 - \frac{\varepsilon}{\delta_1} - \frac{\varepsilon}{\delta_3} - \frac{\varepsilon}{\delta_4} - \frac{\varepsilon}{\delta_5}]\sum_{i=1}^{N}\|\hat Y_i\|^2 - [l_5(1-c) - \frac{\varepsilon}{\delta_2} - \frac{\varepsilon}{\delta_6}]$$
$$\times\sum_{j\in K}\|\hat{\boldsymbol\lambda}_j\|^2 + \frac{\varepsilon c^2 l_1^2\delta_1}{4}\|x-x^*\|^2 + \frac{\varepsilon c^2\kappa^2\delta_2}{4}\|x-x^*\|^2\max\{\|\nabla_x g_j(Y)\|^2\}$$
$$+\frac{\varepsilon c^2 l_2^2\kappa^2\delta_3}{4}\|x-x^*\|^2\max\{\lambda_{1j}^2\} + \frac{\varepsilon\delta_4 c^2 l_3^2}{4}(\sum_{j\in K}\|\lambda_{1j}-\lambda_{1j}^*\|)^2 \tag{10}$$
$$+\frac{\varepsilon\delta_5(1-c)^2}{4}\sum_{i=1}^{N}\left\|\frac{1}{\varepsilon}\frac{dx_i}{dt}\mathbf{1}\right\|^2 + \frac{\varepsilon\delta_6(1-c)^2}{4}\sum_{j\in K}\left\|\frac{1}{\varepsilon}\frac{d\lambda_{1j}}{dt}\mathbf{1}\right\|^2.$$

Let ε be chosen such that $0<\varepsilon<\min\{\varepsilon_1,\varepsilon_2\}$, where $\varepsilon_1 = (\frac{1}{\delta_1}+\frac{1}{\delta_3}+\frac{1}{\delta_4}+\frac{1}{\delta_5}+\frac{1}{\delta_7})^{-1}(1-c)l_4$,

,

and $\varepsilon_2 = (\frac{1}{\delta_2}+\frac{1}{\delta_6}+\frac{1}{\delta_8})^{-1}(1-c)l_5$. Then (10) can be rewritten as

$$\dot V \le -\varepsilon[c\beta(z) + \frac{1}{\delta_7}\sum_{i=1}^{N}\|\hat Y_i\|^2 + \frac{1}{\delta_8}\sum_{j\in K}\|\hat{\boldsymbol\lambda}_j\|^2] + o(\varepsilon\delta_1)$$
$$+o(\varepsilon\delta_2)+o(\varepsilon\delta_3)+o(\varepsilon\delta_4)+o(\varepsilon\delta_5)+o(\varepsilon\delta_6)$$
$$\le -\varepsilon c_\rho\rho(\|\Delta\|) + \varepsilon\sum_{i=1}^{6}o(\delta_i),$$

where $c_\rho = \min\{c,\frac{1}{\delta_7},\frac{1}{\delta_8}\}$ and $\rho(\|\Delta\|) = \beta(z) + \sum_{i=1}^{N}\|\hat Y_i\|^2 + \sum_{j\in K}\|\hat{\boldsymbol\lambda}_j\|^2$.

Therefore, for each $\|\Delta(t)\| \ge \rho^{-1}(\frac{2}{c_\rho}(\sum_{i=1}^{6}o(\delta_i)))$, $\dot V \le -\frac{\varepsilon}{2}c_\rho\rho(\|\Delta\|)$. Based on Theorem 4.18 of [20], there exists δ_i^* such that for any $0<\delta_i<\delta_i^*$, $i=1,\cdots,6$, there exists ε^* such that for any $0<\varepsilon<\varepsilon^*$, there exists $T>0$ and class-KL function ρ_0 [20] such that

$$\|\Delta(t)\| \le \rho_0(\|\Delta(0)\|,t), t\le T, \|\Delta(t)\| \le \alpha_1^{-1}(\alpha_2(\rho^{-1}(\frac{2}{c_\rho}(\sum_{i=1}^{6}o(\delta_i))))), t\ge T.$$

Remark 1 In this paper, we consider that each player has several shared constraints, where each constraint is shared by all the players. Our algorithm can also be used to solve the following problem:

$$\min_{x_i} f_i(x_i, x_{-i}), \text{ such that } g_{ij}(x_i) \le 0,\ j \in \mathrm{K}, \tag{11}$$

where each player has multiple private constraints that depend on it's own action. The algorithm can be designed as follows:

$$\begin{aligned} & y_{ii} = x_i, i \in \mathrm{V}, \dot{y}_{ij} = -w_{ij}\sum_{k=1}^{N} a_{ik}(y_{ij} - y_{kj}), j \in \mathrm{V}/\{i\}, \\ & \dot{x}_i = -\bar{k}_i(\nabla_{x_i} f_i(Y_i) + \sum\nolimits_{j\in \mathrm{K}} \nabla_{x_i} \lambda_{ij} g_{ij}(x_i)), \dot{\lambda}_{ij} = \bar{k}_{ij}\lambda_{ij} g_{ij}(x_i), \end{aligned} \tag{12}$$

where player 1 is selected to calculate the common multipliers λ_{1j} and consensus based control laws are used to broadcast them to all the other players. In (12), $\lambda_{1j}(0) > 0$, $\gamma_{ij} > 0$, and $\bar{k}_i = \varepsilon k_i$, where ε is a small positive constant and $k_i > 0$ is a positive constant.

Remark 2 The objective functions in this paper are assumed to belong to class-C^2. In practical applications, there are many objective functions that don't satisfy this condition, e.g., locally Lipchitz objective functions. The proof for locally Lipschitz objective functions are more complicated. The reasons are three-fold: 1) The functions are not differentiable at some points. To deal with this problem, subgradients are used for control design, which leads to nonsmooth dynamical systems. 2) For a locally Lipschitz objective function, its subdifferential is not locally Lipschitz but upper semicontinuous (see [21]), which brings challenges to singular perturbation analysis, the main analysis method used in this paper. 3) The existence and uniqueness of the normalized Nash equilibrium for games with locally Lipschitz objective functions must be analyzed. Here, we only give the algorithm design, and the convergence analysis for the differential inclusion will be our future work. The designed algorithm for locally Lipschitz objective functions is described as follows:

$$\begin{aligned} & \dot{x}_i = -\bar{k}_i(h_i(Y_i) + \sum\nolimits_{j\in \mathrm{K}} \lambda_{ij} \nabla_{x_i} g_j(Y_i)), \\ & \dot{\lambda}_{1j} = \bar{k}_{1j}\lambda_{1j} g_j(Y_1), \dot{\lambda}_{ij} = -\gamma_{ij}\sum_{k=1}^{N} a_{ik}(\lambda_{ij} - \lambda_{kj}), j \in \mathrm{K}, i \ne 1, \end{aligned} \tag{13}$$

where Y_i is produced by (4). In (13), $h_i(Y_i)$ is the subgradient of $f_i(Y_i)$, defined as $h_i(Y_i) \in \partial_{x_i} f_i(Y_i)$, where $\partial_{x_i} f_i(Y_i)$ is the subdifferential of $f_i(x)$ at Y_i, and $\bar{k}_i$ and γ_{ij} are defined the same as in (12).

5. Simulation

Example 1 A network of 5 players is shown in Fig. 1. Let $f_1(x,t) = x_1^2 + (x_1 - x_2 - 3)^2$, $f_2(x,t) = (x_1 - x_2)^2 + (x_2 - x_3)^2$, $f_3(x,t) = x_3^2 + (x_2 - x_3)^2$, $f_4(x,t) = (x_4 - x_3 - 1)^2$, and $f_5(x,t) = x_5^2 + (x_5 - x_1)^2$ be the local objective functions for players 1-5, respectively. The constraint set is $\Omega = \{x \in \mathbb{R}^5 \mid \|x - d\|^2 \le 9, d = [1,2,1,0,-1]^T\}$. Let (4) and (5) be the updating law. Fig. 2 shows the simulation result of the estimate on the optimal solutions $x_1^*(t), x_2^*(t),\ \cdots,\ x_5^*(t)$.

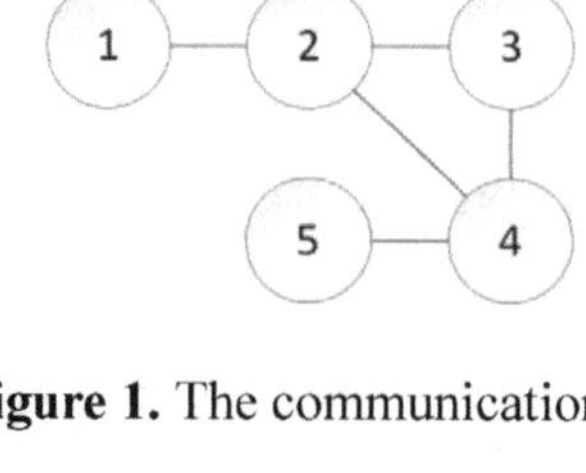

Figure 1. The communication graph.

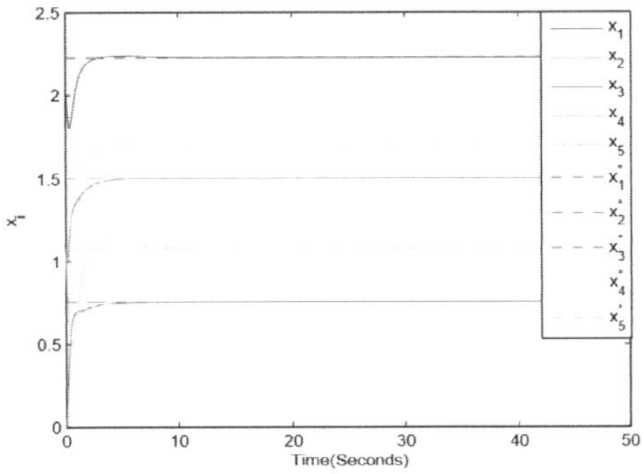

Figure 2. The estimate on the Nash equilibrium.

6. Conclusions

In this paper, we studied the distributed Nash equilibrium seeking problems for generalized convex games with multiple shared constraints. Supposing that each player can exchange information with its neighbors via a connected undirected graph, continuous-time control laws were designed such that each player minimizes its own cost function in a distributed way. In future, we will consider distributed continuous-time Nash equilibrium seeking for nonsmooth objective functions with convergence analysis. In addition, we will relax the assumptions used in the paper.

Acknowledgments
This work was supported by Singapore Economic Development Board under EIRP grant S14-1172-NRF EIRP-IHL.

References

[1] Nash J 1951 Non-cooperative games *Annals of mathematics* 286–295
[2] Ratliff L J, Burden S A and Sastry S S 2013 Characterization and computation of local nash equilibria in continuous games *2013 51st Annual Allerton Conference on Communication, Control, and Computing (Allerton)* pp 917–924
[3] Vamvoudakis K G, Lewis F L and Hudas G R 2012 Multi-agent differential graphical games: Online adaptive learning solution for synchronization with optimality *Automatica* **48** 1598–1611
[4] Frihauf P, Krstic M and Basar T 2012 Nash equilibrium seeking in noncooperative games *IEEE Transactions on Automatic Control* **57** 1192–1207
[5] Ye M and Hu G 2015 Distributed seeking of time-varying Nash equilibrium for non-cooperative games *IEEE Transactions on Automatic Control* **60** 3000–3005
[6] Ye M and Hu G 2016 Distributed extremum seeking for constrained networked optimization and its application to energy consumption control in smart grid *IEEE Transactions on Control Systems Technology* **24** 2048–2058
[7] Rosen J B 1965 Existence and uniqueness of equilibrium points for concave n-person games *Econometrica: Journal of the Econometric Society* 520–534
[8] Pang J S and Fukushima M 2005 Quasi-variational inequalities, generalized Nash equilibria, and multi-leaderfollower games *Computational Management Science* **2** 21–56
[9] Facchinei F and Kanzow C 2007 Generalized Nash equilibrium problems *4OR: A Quarterly Journal of Operations Research* **5** 173–210
[10] Debreu G 1952 A social equilibrium existence theorem *Proceedings of the National Academy of Sciences* **38** 886–893
[11] Arrow K J and Debreu G 1954 Existence of an equilibrium for a competitive economy *Econometrica: Journal of the Econometric Society* 265–290

[12] Facchinei F, Fischer A and Piccialli V 2007 On generalized Nash games and variational inequalities *Operations Research Letters* **35** 159–164
[13] Gharesifard B and Cort´es J 2014 Distributed continuous-time convex optimization on weight-balanced digraphs *IEEE Transactions on Automatic Control* **59** 781–786
[14] Yi P, Hong Y and Liu F 2015 Distributed gradient algorithm for constrained optimization with application to load sharing in power systems *Systems & Control Letters* **83** 45–52
[15] Cherukuri A, Mallada E and Cort´es J 2016 Asymptotic convergence of constrained primal–dual dynamics *Systems & Control Letters* **87** 10–15
[16] Bai L, Ye M, Sun C and Hu G 2016 Distributed control for optimal economic dispatch via saddle point dynamics and consensus algorithms *IEEE 55th Conference on Decision and Control (CDC)* pp 6934–6939
[17] Sun C, Ye M and Hu G 2017 Distributed time-varying quadratic optimization for multiple agents under undirected graphs *IEEE Transactions on Automatic Control* **62** 3687 - 3694
[18] Ye M and Hu G 2017 Distributed Nash equilibrium seeking by a consensus based approach *IEEE Transactions on Automatic Control* **62** 4811 - 4818
[19] Zhu M and Frazzoli E 2016 Distributed robust adaptive equilibrium computation for generalized convex games *Automatica* **63** 82–91
[20] Khalil H K 2002 Nonlinear Systems *New Jewsey, Prentice Hall* **9**
[21] Niederl¨ander S K and Cort´es J 2016 Distributed coordination for nonsmooth convex optimization via saddlepoint dynamics *arXiv preprint arXiv:1606.09298*

Research on Equivalent Tests of Dynamics of On-orbit Soft Contact Technology Based on On-Orbit Experiment Data

F Yang[1]**, Z H Dong** [1]**, X Ye**[2]

[1] School of Space Information, Space Engineering University, 101416 Beijing, China
[2] Department of Graduate Management, Space Engineering University, 101416 Beijing, China

Email: dzh.bj@163.com

Abstract. Currently, space robots have been become a very important means of space on-orbit maintenance and support. Many countries are taking deep research and experiment on this. Because space operation attitude is very complicated, it is difficult to model them in research lab. This paper builds up a complete equivalent experiment framework according to the requirement of proposed space soft-contact technology. Also, this paper carries out flexible multi-body dynamics parameters verification for on-orbit soft-contact mechanism, which combines on-orbit experiment data, the built soft-contact mechanism equivalent model and flexible multi-body dynamics equivalent model that is based on KANE equation. The experiment results approve the correctness of the built on-orbit soft-contact flexible multi-body dynamics.

1. Introduction

With the rapid development of space technology, the on-orbit maintenance and support have become new strategic commanding heights during international competition, which aim to enhance control and support for space environment by implementing all kinds of space operations. The space contact technology is the basic aspect of on-orbit maintenance and support [1,2]. Currently, the major space contact technology includes manipulator, space fly claw, and space docking. And space advanced countries have already started to discover those technologies. The USA [3], Japan [4] and China [5] have already completed on-orbit experiment on space manipulator. However, those experiments were all based on space hard contact technology, i.e. to directly contact objects by rigid components. Because the spacecraft base and space objects should not generate any abnormal attitude disturbance during on-orbit capturing. This greatly limits the improvement of on-orbit operation ability for space contact technique and limits the application range of space objects. In order to solve the problem of hard-contact technology, this paper proposes a new technology called soft-contact technology. The core idea is to construct a flexible space manipulator model and rigid-flexible compound control method which can make the collision momentum controllable buffering and unloading, and can transfer collision momentum to harmonic collision momentum. [6]

As the idea of space soft-contact technology remains at the stage of theory verification, it still requires verifying the feasibility of collision momentum controllable buffering and unloading implemented by this structure. Currently, most of the similar space robots experiment verifications were implemented by simulation. For example, Zhang developed a set of on-orbit service simulation system for space intelligent operation equipment, which adopted HLA/RTI simulation framework, to implement abilities like space situation monitoring, mission planning, manoeuvring planning, observation and control resources computation, orbital manoeuvre simulation, on-orbit operation simulation, control instruction generation and dispatch. [7] But its simulation accuracy depends on the accuracy of modelling, which still has a certain gap between the real cases. Air bearing allows us to take experiment on space arm under the condition of simulating 0-G floating in 2D space. In Japan, ETS-VII satellite tested its control algorithm by air bearing table. [8, 9] While another technique to reduce weight is to use pool to implement neutral buoyancy. It can carry out experiment in 6-DOF space, without time limitation. MIT has carried out many tests for long distance operation robots with NBL of NASA. Besides, gravity can be compensated by suspended system or balance mechanism that is used for space manipulator to reduce gravity test [10]. Bai [11] took research on gravity compensation system. Their experiment applied light cable to go through several chain wheels, and ended with the same counterweight with the same valid weight of simulation robot. The above-mentioned method can simulate on-orbit work status of space manipulators in a certain degree, but the simulation results still have a certain error due to the limitation of objective conditions. This paper makes use of current typical space multi-rigid body manipulator system and its space experiment data to develop the multi-body dynamics on-orbit verification for on-orbit soft-contact mechanism, and provides practice data support for the verification of rigid-flexible compound multi-body dynamics model.

The key point of this paper is how to verify the validity of the proposed space flexible robot dynamics model by current space robot on-orbit testing data. Firstly, we set the dynamics parameters for space flexible robot according to the limitations of current space system. In the meantime, we construct dynamics model based on those parameters. Second, this paper designs an equivalent experiment of flexible multi-body dynamics model according to on-orbit testing data, which compares the experiment data with the results given by KANE dynamics equation of flexible space robot, and corrects flexible characteristic parameters by particle swarm algorithm. Finally, this paper carries out test on the validity of the proposed dynamics model by on-orbit testing data.

The rest of paper is organized as follows, section 2 describes experiment demand analysis and on-orbit equivalent experiment framework. Section 3 explains the construction of on-orbit equivalent experiment model. Section 4 verifies flexible multi-body dynamics parameter and shows simulation results. Section 5 gives a brief conclusion on the paper.

2. Experiment Demand Analysis and Equivalent Experiment Framework

On-orbit soft-contact mechanism should have three basic abilities; the first is the ability of having workspace to get the six surfaces of target to carry out maintenance operations. Second is to have a certain degree of flexibility, which can make sure flexible mechanism to operate target satellites in various configurations, in order to meet the requirement of some on-orbit contact task for flexible special configuration. At last, the path plan process of on-orbit soft-contact mechanism should have lower control complexity, to make sure the stability of the control process.

The biggest difference between on-orbit soft-contact mechanism and current space multi-rigid manipulator is the buffering controlling ability of touching collision momentum. The key to implement this ability is whether the flexible multi-body dynamics model can be built in space environment or not, especially, whether it can accurate construct space object collision dynamics model or not. Thus, it has to discuss about the equivalent experiment, dynamics characteristic parameter and model test and verification of on-orbit dynamics for on-orbit soft-contact mechanism.

On-orbit equivalent experiment and verification of soft-contact mechanism dynamics mainly contain three stages. First, to limit the design of flexible multi-body model in terms of current space multi-body system. Second, to develop the research on equivalent experiment of on-orbit soft-contact mechanism flexible multi-body system. Third, to carry out research on parameter verification of the built flexible multi-body dynamics equations.

The general experiment flow and method are show as figure 1.

3. Construction of On-orbit Equivalent Experiment Model

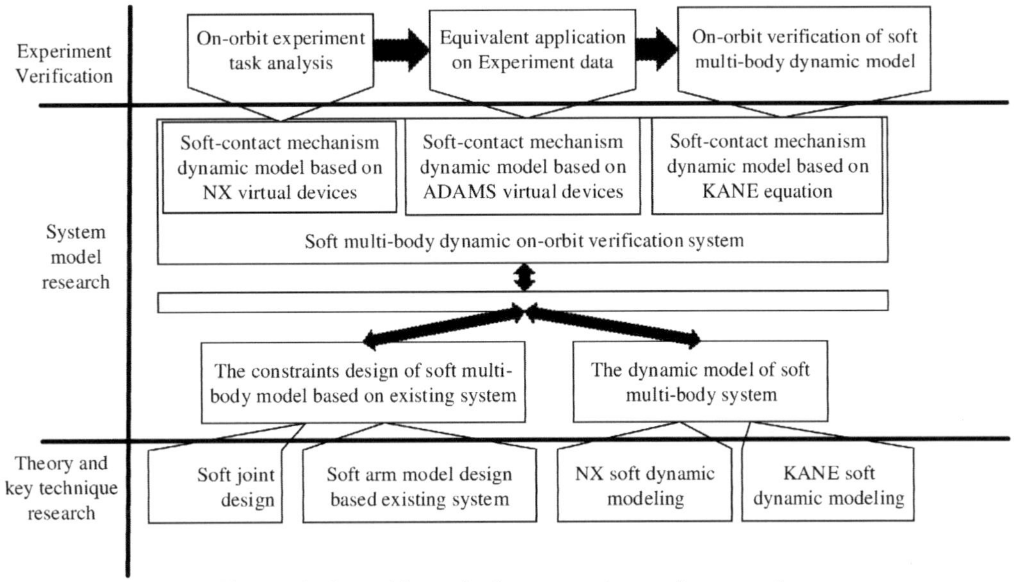

Figure 1. On-orbit equivalent experiment framework.

The basic structure of current on-orbit space multi-rigid manipulator system is consisted of 3 same unified double joints and carbon fibre arms, which has 6 degrees of freedom, and the system basic structure. This paper constructs equivalent model of flexible angular momentum and linear momentum controllable joints of on-orbit soft-contact mechanism by the equivalent design of arms and joints, in terms of the flexible multi-body dynamics characteristics when current space manipulator system's joint harmonic drive mechanism is under the condition of deadlock, at the moment of capturing space objects.

Based on current space multi-rigid manipulator system, on-orbit soft-contact mechanism does not only have rigidity parameters, but also has flexible joint parameters. Take the space multi-rigid manipulator shown as figure 2 as an example, the flexibility of its joint harmonic drive mechanism under the condition of deadlock can be equivalent to on-orbit soft-contact flexible and controllable joints which have a certain torsional rigidity. Thus, equivalent on-orbit soft-contact mechanism can be treated as a space manipulator system which has 6 DOF flexible joints.

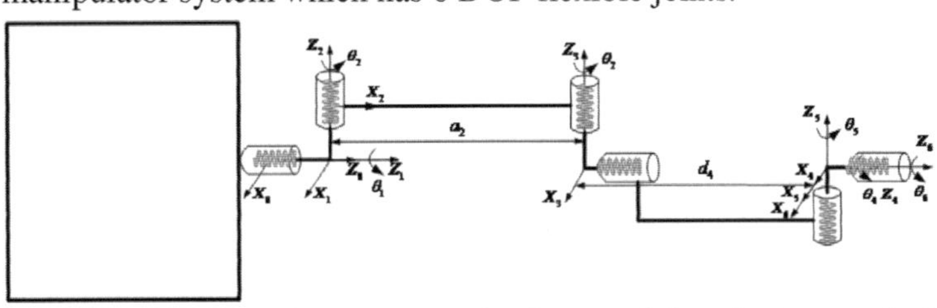

Figure 2. On-orbit soft-contact mechanism model equivalent analysis.

According to the above equivalent model design, the flexible angular momentum/ linear momentum joints can be simplified to torsional spring/spring and damping system, and arms are simplified into rigid-bodies, when doing dynamics modelling on on-orbit soft-contact mechanism. Thus, on-orbit soft-contact mechanism dynamics model can be dispersed to multi-rigid body model that is connected with torsional spring/spring and damping system. The satellite base has six DOFs, which can be considered as a virtual arm which has 6 DOFs. On the basis of the above discussion, according to the construction method of KANE dynamics equation, the floating manipulators which have N joints and 6N DOFs can be considered as a fixed base manipulator system which has (6N+6) DOFs, and also the general dynamics equivalent equation can be constructed as following,

$$\{\dot{y}_n\} = \left[a_{l\eta}\right]^{-1} \cdot \{f_l\} \tag{1}$$

Where $\dot{y}_\eta$ is the derivative of generalized speed;

$$\{\boldsymbol{f}_l\}=\{\boldsymbol{F}_l\}-\sum_{k=1}^{N+1}\boldsymbol{v}_{kl}\cdot\boldsymbol{m}_k\cdot\dot{\boldsymbol{V}}_{k\eta}\cdot\{\boldsymbol{y}_\eta\}-\sum_{k=1}^{N+1}\omega_{kl}\cdot[\omega_k^{\mathbf{R}}\times(\boldsymbol{I}_k\cdot\omega_k^{\mathbf{R}})]-\sum_{k=1}^{N+1}\omega_{kl}\cdot\boldsymbol{I}_{kc}\cdot\dot{\boldsymbol{W}}_{k\eta}\cdot\{\boldsymbol{y}_\eta\} \quad (2)$$

$$[\boldsymbol{a}_{l\eta}]=\sum_{k=1}^{N+1}\boldsymbol{v}_{kl}\cdot\boldsymbol{m}_k\cdot\boldsymbol{V}_{k\eta}+\sum_{k=1}^{N}\omega_{kl}\cdot\boldsymbol{I}_{kc}\cdot\omega_{k\eta} \quad (3)$$

Equation (1) is non-linear simultaneous differential equations, K_k is the corresponding spring rigidity coefficient, i.e. the flexible joint rigidity parameter which is required to be verified in the equivalent model of on-orbit soft-contact mechanism. The above equations contain 6N+6 generalized coordinates and 6N+6 generalized speed, i.e. 12N+12 unknown variables in total [12].

4. On-orbit Verification of Flexible Multi-body Dynamics Parameter

The docking process of current space manipulator end execution device to space objects is as follows:

(1) By controlling the relative speed of two docking satellites, to make space manipulator end docking mechanism to form a docking envelope area for object satellites.

(2) To gradually close the end docking mechanism, to make the collision momentum between the two docking satellites reach rigid connection, and do not cause too much collision momentum to satellite base which carries manipulator. During the process of capturing object satellite, two docking satellite bases stay in floating state.

After analysis the collision dynamics response data of space arm system for the two satellites in the docking process, we can obtain important dynamics response characteristics from the docking process of manipulator system, which can provide on-orbit experiment data support for developing multi-body dynamics characteristic test verification of on-orbit soft-contact mechanism.

We use the current space manipulator system on-orbit data to develop flexible multi-body dynamics experiment, parameter verification and test verification based on KANE equation. More details about the flow chart is shown as figure 3.

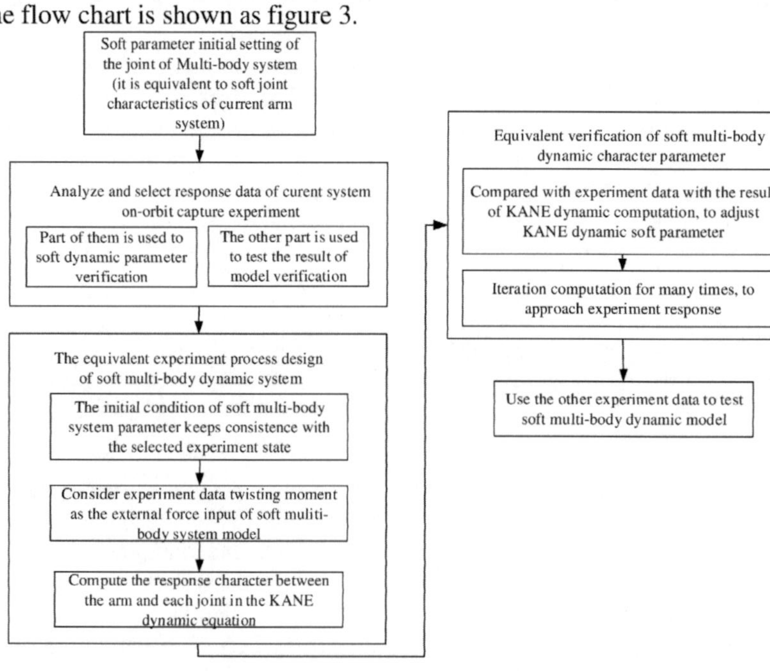

Figure 3. Flow of flexible multi-body dynamics parameter verification of soft-contact mechanism.

Step 1: Acquire of equivalent parameters of on-orbit soft-contact mechanism flexible joints.

Making the twisting stiffness of the harmonic reducer of current space manipulator system as the basic parameter of flexible multi-body dynamics in KANE equation of flexible multi-body system.

Step 2: Analyse and select dynamics response data of space experiment of on-orbit space manipulator system.

On-orbit experiment data is divided into two parts. One part is used for flexible dynamics parameter verification, and the other part is used to test the results of model verification.

Step 3: The equivalent experiment process design of on-orbit soft-contact mechanism flexible multi-body dynamics.

Step 4: Comparing the experiment data of current space manipulator system with the results of KANE dynamics equation. And using their differences to correct the flexible characteristic parameter of KANE dynamics equation.

Step 5: Using the other experiment data of on-orbit contacts experiment test the correctness and effectiveness of the verified soft multi-body dynamics model.

The location of collision point, and joint angular changes between no collision and the moment of collision can be obtained by on-orbit data during the docking process between two satellites. The collision force of space manipulator end can be obtained by Jacobi equation. The angle of each joint response of space manipulator can be obtained by corresponding collision point on-orbit data, and Euler angle can be got by coordinate transformation equation. Then, bring the end collision force of above mentioned collision points and corresponding joint response Euler angles to equation (1) for solution. Finally, comprehensively make use of particle swarm algorithm and damper positive-inverse model, to implement the momentum control of on-orbit soft-contact mechanism, more detailed flow is show as follows,

(1) When the end of on-orbit soft-contact mechanism is collided, the displacement $x(t)$ of damper at t moment can be tested by motion sensor.

(2) Joint damper displacement $x(t)$ at t moment, particle swarm algorithm, and KANE dynamics model can be used to get optimized damping force at t moment by iteration.

(3) We use damper positive-inverse model to output the expected optimized damping force $\boldsymbol{u}(t)$ to act on flexible controller of multi-arm joints. And then to obtain damper displacement $x(t+1)$ of each joint at t+1 moment tested by sensor. We reuse step 2 to compute the optimized damping force at t+1 moment, and get the torsional rigidity, and we repeat the same procedure again and again.

We use the above verified torsional rigidity of harmonic reducer as the equivalent flexible parameter for on-orbit soft-contact mechanism dynamics equations, and bring other collision sample data obtained by space manipulator on-orbit experiment to the equivalent KANE equation of on-orbit soft-contact mechanism, and obtain the angle of each joint in the collision process, to compare with the response data of each corresponding on-orbit joint, so as to verify the correctness of the built flexible multi-body dynamics of on-orbit soft-contact mechanism. After the flexible joint parameter verification of on-orbit soft-contact mechanism dynamics and the test of current on-orbit space manipulator system, the results are as figure 4-7. The dash lines represent the joint response original data at the collision moment in on-orbit experiment, and the solid lines represent the response data of flexible multi-body equivalent model joint at the collision moment of on-orbit soft-contact mechanism. The tendency of two lines are almost the same. The equivalent experiment comparison results show that compared with current on-orbit experiment data, the accuracy error of on-orbit soft-contact mechanism flexible multi-body dynamics model has been controlled to within 35% after being verified.

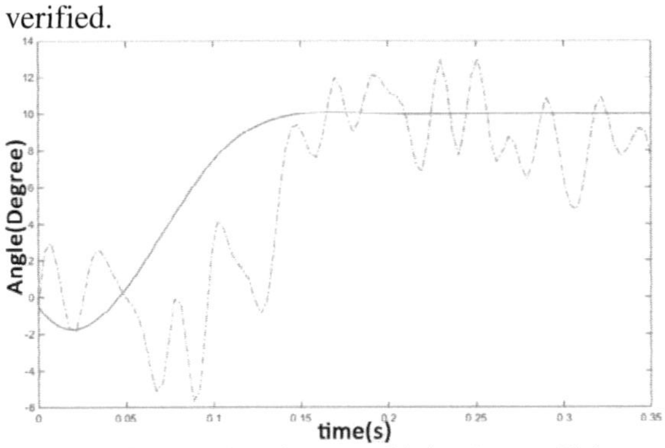

Figure 4. Rotation degree of joint 1 at collision moment.

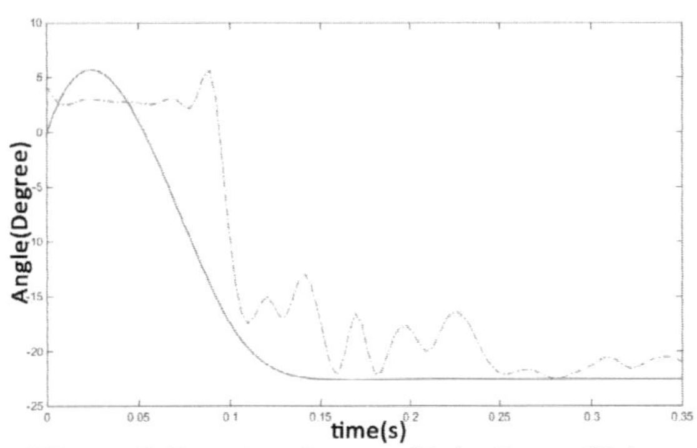

Figure 5. Rotation degree of joint 2 at collision moment.

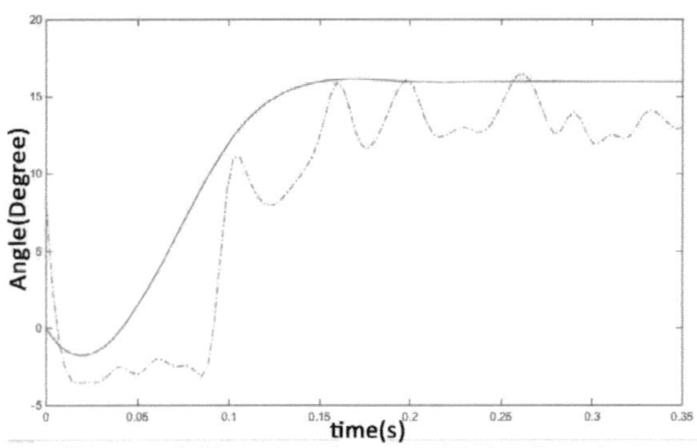

Figure 6. Rotation degree of joint 3 at the collision moment.

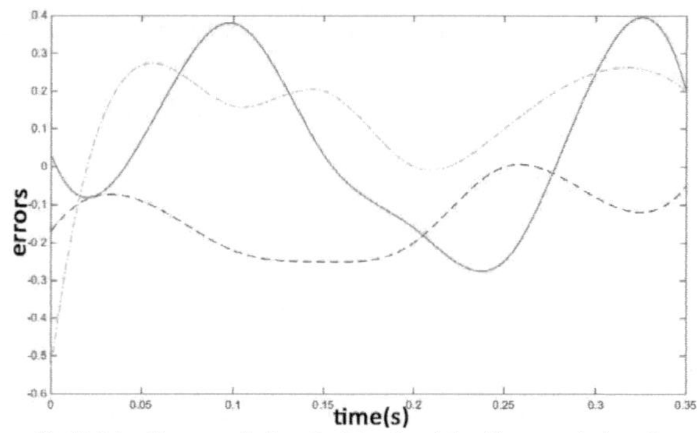

Solid indicates joint 1, Dashed indicates joint 2 and chain dotted line indicates joint 3.

Figure 7. Error verification of Equivalent model

5. Conclusion

This paper firstly develops equivalent experiment requirement analysis and equivalent experiment basic framework and its contents based on current space manipulator system. Secondly, this paper implements space flexible multi-body dynamics modelling based on current space manipulator system, which mainly focuses on the building up of space flexible multi-body dynamics equivalent model based on KANE equation. Finally, according to the above research, this paper develops equivalent parameter verification using on-orbit data, and the results show that the correctness of the built up on-orbit soft-contact mechanism flexible multi-body dynamics.

References

[1] Richard W M 2000 Micro-satellite based, On-orbit servicing work at the air force research *Aerospace Conference Proceedings* (Big Sky: IEEE) vol 4 pp 215-26

[2] Naigang C, Ping W, Jifeng G and Xing C 2007 A Review of On-Orbit Servicing *Journal of Astronautics* **28** 33-39

[3] Stamm S, Pejmun M 2004 Orbital Express Contact System: concept to reality *In Spacecraft Platforms and Infrastructure Proc. of SPIE* (Bellingham: SPIE) vol 5419 pp78-91

[4] Oda M 1999 Space robot experiments on NASDA's ETS-VII satellite-preliminary overview of the experiment results *In Proc. Of IEEE International Conf. on Robotics and Automation* (Detroit: IEEE) vol 2, 1390-1395.

[5] Ying Z 2013 *Modeling and Control Based on the Single Joint Locked Space Manipulator* (Chongqing: Chongqing University Press)

[6] Xin Y, Zhenghong D, Jiacai H and Fan Y 2016 Research on kinematic characteristic of a new flexible space manipulator based on the integrated double joints *In Proc.of 2016 5th International Conf. on Mechatronics and Control Engineering* (Venice: ACM) 86-91

[7] Xuebo Z, Xiaohong Y and Zhihai Z 2014 The Modeling and Simulation Research of the Space Intelligent Operation Equipment's In-orbit Service *Journal of Equipment Academy* **25** 47-51

[8] Umetani Y and Yoshida K 1989 Experimental study on two-dimensional free-flying robot satellite model. *In NASA conf. on space telerobotics* (Pasadena: JPL) vol 5pp 215–24

[9] Yoshida K 1995 Experimental study on the dynamics and control of a space robot with experimental free-floating robot satellite (EFFORTS) simulators *Advanced Robotics* **9** 583–602

[10] Lu Q, Ortega C and Ma O 2011 Passive gravity compensation mechanisms: technologies and applications *Recent Patents on Engineering* **5** 32–44

[11] White G and Xu Y 1994 An active vertical-direction gravity compensation system *IEEE Transactions on Instrumentation and Measurement* **43** 786–92.

[12] Zhenghong D, Xin Y, Jiacai H and Zhiwei L 2017 Research on Dynamic Modeling of Soft-Contact Technology Based on Kane's Equations *Proc. of the 2017 18th Int. Conf. on Advanced Robotics* (Hongkong: IEEE) 353-58

IOP Conf. Series: Journal of Physics: Conf. Series **1016** (2018) 012014 doi:10.1088/1742-6596/1016/1/012014

Research on The Construction of Flexible Multi-body Dynamics Model based on Virtual Components

Z H Dong [1], X Ye[2], F Yang[1]

[1] School of Space Information, Space Engineering University, 101416 Beijing, China
[2] Department of Graduate Management, Space Engineering University, 101416 Beijing, China

Email: dzh.bj@163.com

Abstract. Focus on the harsh operation condition of space manipulator, which cannot afford relative large collision momentum, this paper proposes a new concept and technology, called soft-contact technology. In order to solve the problem of collision dynamics of flexible multi-body system caused by this technology, this paper also proposes the concepts of virtual components and virtual hinges, and constructs flexible dynamic model based on virtual components, and also studies on its solutions. On this basis, this paper uses NX to carry out model and comparison simulation for space manipulator in 3 different modes. The results show that using the model of multi-rigid body + flexible body hinge + controllable damping can make effective control on amplitude for the force and torque caused by target satellite collision.

1. Introduction

With the rapid development of space technology, space on-orbit maintenance and support have become the hottest topic among different countries, which aim to enhance control and support for space environment by implementing all kinds of space operations. The space capture technology is an basic aspect of on-orbit maintenance and support. Currently, the world's major space powers have already completed space contact technology on-orbit tests, such as NASA's “Orbital Express” [1] and Japan's ETS-VII [2]. However, these tests basically adopt "hard-contact technology", that is, directly contacting with rigid components in orbit capture technology. This technology’s space target operating condition is very harsh. The main reason is that the main spacecraft’s space capturing device is generally composed of a single rigid arm or multi-rigid arm cascade, and the space target relative attitude requires keeping highly stability during operating. This requires that any abnormal disturbance should not happen when the main spacecraft and space targets are contacting with each other, otherwise a great impact collision momentum may be generated. This greatly limits the space contact technology to improve on-orbit operational capacity and the range of applications of space target.

To resolve the problem of on-orbit hard-contact technology, this paper proposes a new concept called soft-contact technology. Its core idea is to build up a soft-capture mechanism model and rigid-flexible compound control method, in order to achieve transformation from stepped collision momentum to harmonic attenuation momentum. And the key parameters, such as harmonic amplitude, phase and convergence time, can be controlled according to the desired function. This idea can make collision momentum controlled buffering and unloading come true. [3]

On-orbit soft-contact mechanism constructed by this idea is a complex multi-flexible body system consisted of multiple joints and multiple arms, which has multiple inputs and multiple outputs, and they have coupling relations in between with strong non-linear characteristic in the meantime. This research on flexible multi-body system collision dynamics of on-orbit soft-contact mechanism aims to design the control strategy. The key to solve flexible multi-body system collision dynamics is to build accurate rigid-flexible compound dynamics model and correctly deal with collision process, and then find out effective and stable solutions. According to the different hypotheses of collision process, the multi-body system collision dynamics model methods can be classified into impulse momentum method, continuous contacting force method, contact constraint method, etc. [4] The above mention methods have their own advantages and limitations. Zhang etc. brought the idea of impulse potential energy, and obtained the collision dynamics equation out of flexible-hinge rigid-arm robot system, which can calculate the mutations of system general velocity after collision [5]. However, this method has certain constraints in the research on complex flexible multi-body system. Fang adopted contacting theory and non-linear damping theory to build contacting collision model, and exported flexible variables and rigid-flexible compound dynamics equation of collision. [6] The biggest problem of this method is how to select collision model and how to recognize the parameters in the selected model. Han adopted contacting constraint to process collision process, and proposed multi-variable method for the problem of flexible multi-body system collision. In the stage of collision, use floating reference system method to solve the problem in non-collision area, and use non-linear finite element method to solve the problem in the local area of collision. [7] Wang built attached contacting constraint according to the idea of objects that are not embedded in each other, unified dynamics equation of flexible multi-body dynamics before, during, and after collision, brought collision testing mechanism, and implemented the auto changes of topology model. [8] However, similar to the hinge constraint equation in multi-body system, contacting constraint equation and dynamics equation have constructed differential algebra equations, which lead to the difficulty in numeric.

This paper mainly introduces a solution of collision rigid-flexible compound dynamics under Cartesian Coordinate System. On the basis of using contacting constraint method, this paper constructs flexible dynamics model based on virtual components, by bringing the idea of virtual components and virtual hinges. The proposed method makes flexible components without the limitations of all motion conditions except flexible hinge, which reduces the effort to process attached flexible body algorithm. This paper carries out modeling and comparison simulation of three different modes by NX, including multi-rigid body, multi-rigid body + flexible hinge, and multi-rigid body + flexible hinge + controllable damping, according to the proposed idea.

2. Flexible dynamics model based on virtual components

The algorithm structure of flexible multi-body dynamics is shown as figure 1. The old hinge module and force module are only suitable to rigid body, which can be processed to a system consisted of rigid body and flexible body by adding a flexible hinge in the hinge base.

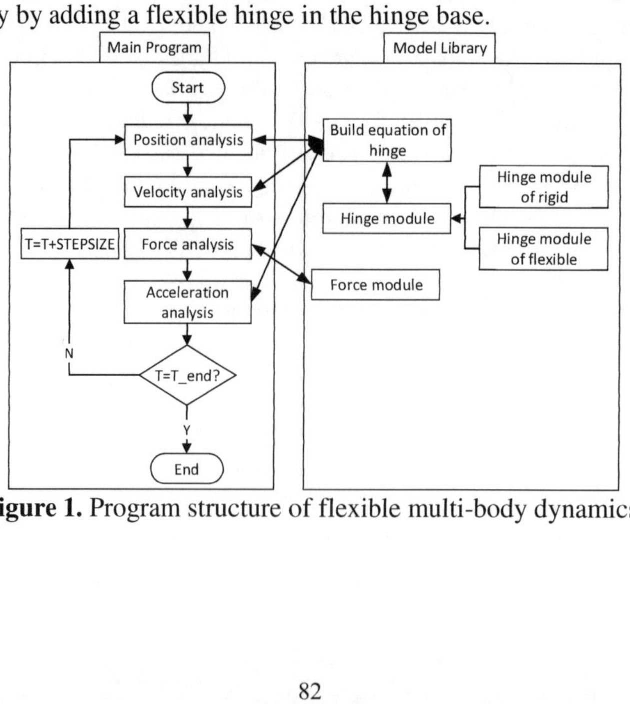

Figure 1. Program structure of flexible multi-body dynamics.

The virtual component connects with the original flexible body through the virtual hinge as shown in Figure 2.

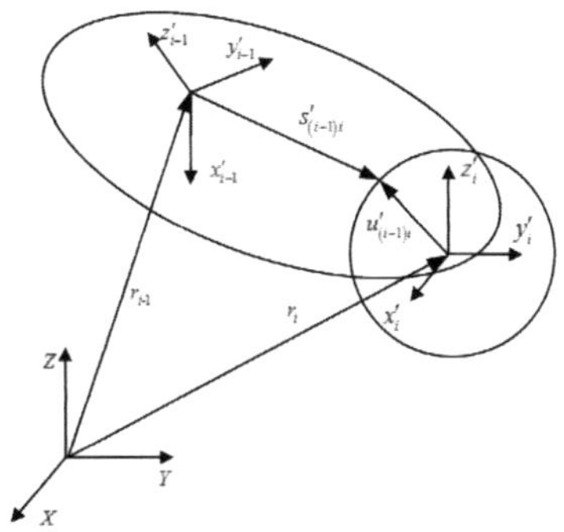

Figure 2. The flexible hinge between flexible body and virtual components.

$s'_{0(i-1)i}$ and $u'_{(i-1)i}$ are unreformed position vector and deformed vector of virtual component origin point relative to flexible reference system. Assuming $\tilde{s}_{(i-1)i} = s'_{0(i-1)i} + u'_{(i-1)i}$, the wave symbol represents oblique symmetric matrix, which can get the velocity recursive equation of flexible hinge as follows,

$$\boldsymbol{Y}_i = \boldsymbol{B}^f_{(i-1)i1}\boldsymbol{Y}_{i-1} + \boldsymbol{B}^f_{(i-1)i2}\dot{\boldsymbol{q}}^f_{(i-1)i} \tag{1}$$

In (1), The superscript f of $\dot{q}^f_{(i-1)i}$ is the modal coordinate vector, and subscript i and $i-1$ are the generalized coordinate in reference system.

$$\boldsymbol{B}^f_{(i-1)i1} = \begin{bmatrix} \boldsymbol{A}^T_{(i-1)i} & -\boldsymbol{A}^T_{(i-1)i}\tilde{s}_{(i-1)i} \\ 0 & \boldsymbol{A}^T_{(i-1)i} \end{bmatrix}, \boldsymbol{B}^f_{(i-1)i2} = \begin{bmatrix} \boldsymbol{A}^T_{(i-1)i}\Phi^R_{i-1} \\ \boldsymbol{A}^T_{(i-1)i}\Phi^\theta_{i-1} \end{bmatrix} \tag{2}$$

Φ^R_{i-1} is modal matrix, which columns is consisted of translation modal vibration mode. Matrix $\boldsymbol{B}^f_{(i-1)i1}$ and $\boldsymbol{B}^f_{(i-1)i2}$ are only the modal coordinate function of flexible body $i-1$. $A^T_{(i-1)i}$ is the attitude matrix of flexible body reference system. (1) defines the kinematics relation between an inner flexible body and an external flexible body, which relation can be deduced. Similarly, the virtual displacement recursive relation between a flexible body and a virtual component is as follows,

$$\boldsymbol{Y}_i = \boldsymbol{B}^r_{(i-1)i1}\boldsymbol{Y}_{i-1} + \boldsymbol{B}^r_{(i-1)i2}\boldsymbol{q}^f_{(i-1)i} \tag{3}$$

In (3),

$$\boldsymbol{B}^r_{(i-1)i1} = \begin{bmatrix} \boldsymbol{A}^T_{(i-1)i} & \tilde{s}'_{(i-1)i}\boldsymbol{A}^T_{(i-1)i} \\ 0 & \boldsymbol{A}^T_{(i-1)i} \end{bmatrix}, \boldsymbol{B}^r_{(i-1)i2} = \begin{bmatrix} -\tilde{s}'_{(i-1)i}\Phi^\theta_{i-1}\text{-}\Phi^R_{i-1} \\ \boldsymbol{A}^T_{(i-1)i}\Phi^\theta_{i-1} \end{bmatrix} \tag{4}$$

The recursive relation of rigid hinge velocity between two rigid bodies can be deduced by (5).

$$\boldsymbol{Y}_i = \boldsymbol{B}^r_{(i-1)i1}\boldsymbol{Y}_{i-1} + \boldsymbol{B}^r_{(i-1)i2}\dot{\boldsymbol{q}}^r_{(i-1)i} \tag{5}$$

In (5), the superscript r is the generalized coordinate of rigid hinge.

$$\boldsymbol{B}^r_{(i-1)i1} = \begin{bmatrix} \boldsymbol{A}^T_{(i-1)i} & -\boldsymbol{A}^T_{(i-1)i}\tilde{s}'_{(i-1)i} + \tilde{\boldsymbol{d}}'_{(i-1)i} - \boldsymbol{A}_{(i-1)i}\tilde{s}'_{(i-1)i}\boldsymbol{A}^T_{(i-1)i} \\ 0 & \boldsymbol{A}^T_{(i-1)i} \end{bmatrix},$$
$$\boldsymbol{B}^r_{(i-1)i2} = \begin{bmatrix} \boldsymbol{A}^T_{(i-1)i}\left(\left(\tilde{\boldsymbol{d}}'_{(i-1)i}\right)_{q(i-1)i}\right) + \boldsymbol{A}^T_{(i-1)i}\tilde{s}'_{(i-1)i}\boldsymbol{A}^T_{(i-1)i}\boldsymbol{H}'_{(i-1)i} \\ \boldsymbol{A}^T_{(i-1)i}\boldsymbol{H}'_{(i-1)i} \end{bmatrix} \tag{6}$$

In (6), $\boldsymbol{H}'_{(i-1)i}$ is fixed by rotation axis. Matrix B is the function of $\dot{q}^r_{(i-1)i}$.

Constraint mechanical system dynamics equation can be obtained by velocity transfer method.

$$\boldsymbol{F} = \boldsymbol{B}^T\left(\boldsymbol{M}\bar{\boldsymbol{T}} + \Phi\boldsymbol{T}/\left(\boldsymbol{Z}^\lambda - \bar{\boldsymbol{Q}}\right)\right) \tag{7}$$

In (7), Φ and λ represent the Lagrange multipliers of cut hinge, $\boldsymbol{M}$ is the mass matrix, and $\bar{\boldsymbol{Q}}$ is force vector, which includes external force, force generated by strain energy and the force caused by velocity. Differential algebra equation (DAE) is consisted of motion equation, constraint equation, $\dot{\boldsymbol{q}} = \boldsymbol{v}$ and $\dot{\boldsymbol{v}} = \boldsymbol{a}$. For the DAE, the following non-linear equation system can be obtained by "cutting space method".

$$\begin{aligned} \boldsymbol{H}(\boldsymbol{p}_n) = \big[& \boldsymbol{U}_0^T(\boldsymbol{q}_n + \beta_0 \boldsymbol{v}_n + \beta_1) \quad \boldsymbol{U}_0^T(\boldsymbol{q}_n + \beta_0 \boldsymbol{a}_n + \beta_2) \quad \boldsymbol{F}(\boldsymbol{q}_n, \boldsymbol{v}_n, \boldsymbol{a}_n, \lambda_n, \boldsymbol{t}_n) \\ & \Phi(\boldsymbol{q}_n, \boldsymbol{t}_n) \quad \dot{\Phi}(\boldsymbol{q}_n, \boldsymbol{v}_n, \boldsymbol{t}_n) \quad \ddot{\Phi}(\boldsymbol{q}_n, \boldsymbol{v}_n, \boldsymbol{z}_n, \boldsymbol{t}_n) \big]^T \end{aligned} \tag{8}$$

In (8), $\boldsymbol{p}_n^T = \left[\boldsymbol{q}_n^T, \boldsymbol{v}_n^T, \boldsymbol{a}_n^T, \lambda_n^T \right]$, β_0, β_1 and β_2 are decided by the coefficients of Fractional Backward Differential Method (BDF) method. $\boldsymbol{U}_0$ is selected to make sure the augmented matrix $\left[\boldsymbol{U}_0^T \quad \Phi_q \right]^T$ is a nonsingular matrix. The given non-linear system can be solved by Newton method.

3. Simulation computation and verification

The dynamic model based on virtual components contains two parts which are multi-rigid body dynamics modeling and rigid-flexible compound body dynamics modeling. Multi-rigid body dynamics modeling considers all components as rigid units. While rigid-flexible compound body dynamics modeling analyzes flexible multi-body dynamics of virtual components and hinges. The dynamics modeling framework is shown as figure 3. The modeling scheme is shown as Table 1.

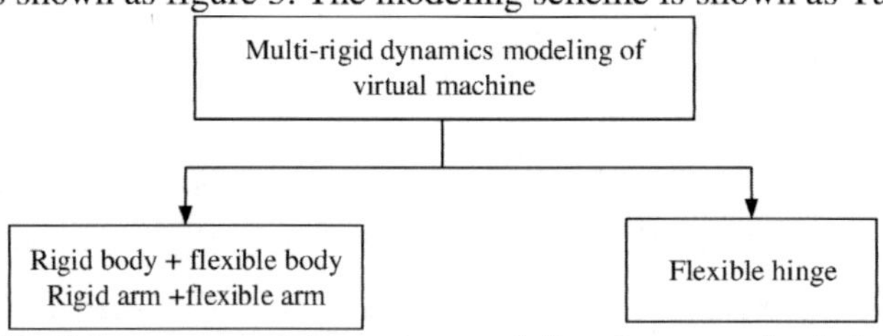

Figure 3. Rigid-flexible compound dynamics modeling framework of virtual components.

Table 1. Modeling scheme.

Model	Explanation
Model 1: Multi-rigid body	All rigid connections between base and joints, and among joints
Model 2: Multi-rigid body + flexible hinge	Based on model 1, add flexible joints
Model 3: Multi-rigid body + flexible hinge + controllable damping	Based on model 2, add controllable damping

We build up a three joints model of on-orbit soft-contact mechanism in NX software, shown as figure 4.

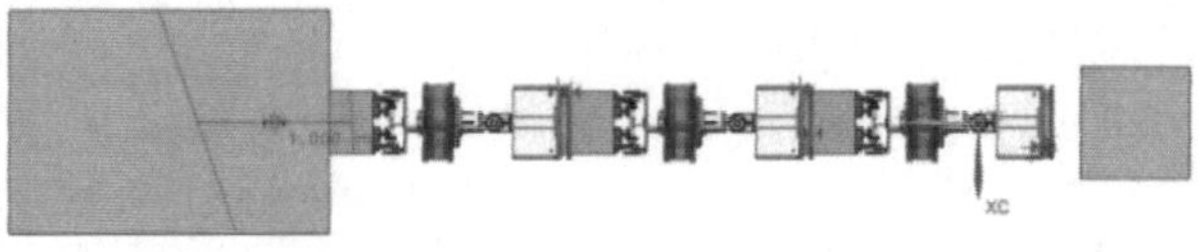

Figure 4. Three joints model.

Assuming the typical collision between space targets and on-orbit soft-contact mechanism has 3 situations as follows,

Collision 1: space targets collide with manipulator with the linear velocity of 1250mm/s along the direction of Z axis.

Collision 2: space targets collide with manipulator with the linear velocity of 1250mm/s and angular velocity of 90deg/s along the direction of Z axis.

Collision 3: space targets collide with manipulator with the linear velocity of 1250mm/s and angular velocity of 90deg/s along the direction of oblique 45 degree.

We carry out simulation analysis on model 1, 2, 3 in table 1 in terms of the above collision situations. We mainly compare the disturbance of rigid-flexible dynamics characteristics of soft-contact mechanism on satellite base, under the same collision conditions.

(1) The disturbance dynamics simulation analysis caused by satellite base under the condition of collision 1 (the target satellites collide with the linear velocity along the direction of Z axis) are shown as figure 5.

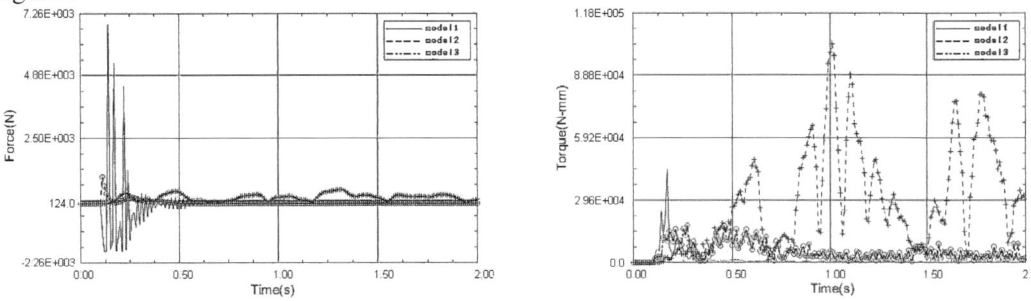

(a) Changes of satellite base disturbance force. (b) Changes of satellite base disturbance torque.

Figure 5. The changing curve of disturbance force and disturbance torque for satellite base.

(2) The disturbance dynamics simulation analysis caused by satellite base under the condition of collision 2 (the target satellites collide with the linear velocity + angular velocity along the direction of Z axis) are shown as figure 6.

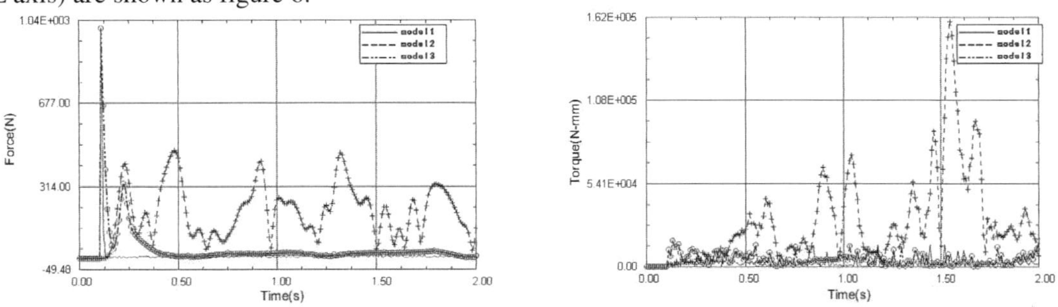

(a) Changes of disturbance force for base. (b) Changes of disturbance torque for base.

Figure 6. The changes of disturbance force and disturbance torque for base.

(3) The disturbance dynamics simulation analysis caused by satellite base under the condition of collision 3 (the target satellites collide with the linear velocity + angular velocity along the direction of oblique 45 degree) are shown as figure 7.

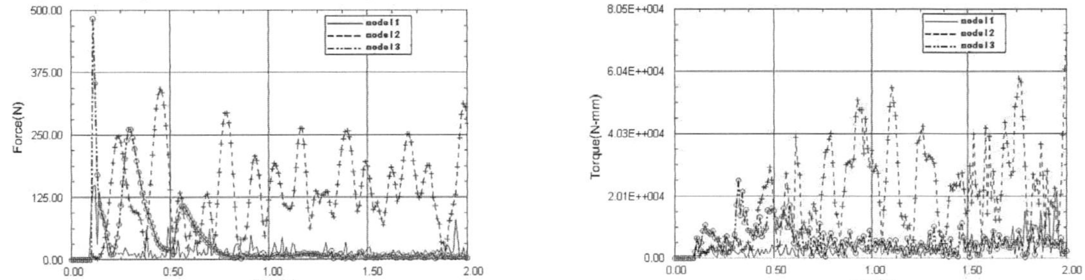

(a) Disturbance force changing curve of base. (b) Disturbance torque changing curve of base.

Figure 7. The changes curve of disturbance force and disturbance torque of base.

From figure 5 to 7, in the aspect of disturbance force and torque acted on satellite base, the dynamics simulation results of 3 models are as follows. On the starting moment, compared with model 2, 3, model 1 has several continuous pulsed large amplitude disturbance force, followed by model 3, and the force of model 2 is the smallest. In the following stage of disturbance force, model 3 first implements stable convergence on disturbance force, and the amplitude is the smallest, followed by model 1, while model 2 appears small amplitude of shock state. For the disturbance torque of satellite base, model 1, 2, 3 appear different pulsed peaks, where the peak value of model 2 is the largest, followed by model 1, and the peak value of model 3 is the smallest. In the following stage of disturbance torque, the amplitude of model 1 is quickly stabilized, model 3 appears micro shock, and model 2 appears several big continuous peaks of amplitude. Thus, under the condition of collision 1, model 2 has smallest disturbance force peak on satellite base, followed by model 3, and the model 1 is the largest. And model 3 has the smallest disturbance torque peak on satellite base, followed by model 1, and model 2 is the biggest.

4. Conclusion

This paper involves the idea of virtual components, which use the idea of contacting constraints to build a flexible dynamics model, and takes research on its solution. On this basis, this paper uses NX software to implement flexible dynamics model based on virtual components, and carries out comparison simulation analysis on 3 different model of multi-body, multi-body + flexible hinge, multi-body + flexible hinge + controllable damping. The analysis results show that using the mode of multi-body + flexible hinge + controllable damping can make effective amplitude control on the force and torque caused by target satellites.

References

[1] Stamm S, Pejmun M 2004 Orbital Express Capture System: concept to reality *In Spacecraft Platforms and Infrastructure Proc. of SPIE* (Bellingham: SPIE) vol 5419 pp78-91

[2] Oda M 1999 Space robot experiments on NASDA's ETS-VII satellite-preliminary overview of the experiment results *In Proc. Of IEEE Int. Conf. on Robotics and Automation* (Detroit: IEEE) vol 2, 1390-1395.

[3] Xin Y, Zhenghong D, Jiacai H and Fan Y 2016 Research on kinematic characteristic of a new flexible space manipulator based on the integrated double joints *In Proc.of 2016 5th Int. Conf. on Mechatronics and Control Engineering* (Venice: ACM) 86-91

[4] Yuechen D 2012 *Research on the Rigid-Flexible Coupling Impact Dynamics of Flexible Multibody Systems* (Nanjing: Nanjing University of Science & Technology Press)

[5] Dingguo Z and Jorge A 2005 Impact dynamics of flexible-joint robots *Computers & Structures* **38** 25-33

[6] Jianshi F, Baoyu L and Dingguo Z 2008 Continuous Force Approach for Impact Dynamics of Flexible Beam in Large Overall Motion *Journal of Nanjing University of Science and Technology (Natural Science)* **32** 661-65

[7] Shilei H and Jiazhen H 2011 Multi-variable Method for Flexible Multi-body Systems with Contact/Impact *Chinese Journal of Theoretical and Applied Mechanics* **43** 886-93

[8] Dongting W, Jiazhen H and Taihui W 2011 Additional Contact Constraint Method in Impact Stage of Planar Flexible Multi-body Dynamics *Chinese Journal of Theoretical and Applied Mechanics* **43** 1157-61

Simulation study on dynamics model of two kinds of on-orbit soft-contact mechanism

X Ye[1], Z H Dong[2] and F Yang[2]

[1] Department of Graduate Management, Space Engineering University, 101416 Beijing, China
[2] School of Space Information, Space Engineering University, 101416 Beijing, China

Email: dzh.bj@163.com

Abstract. Aiming at the problem that the operating conditions of the space manipulator is harsh and the space manipulator could not bear the large collision momentum, this paper presents a new concept and technical method, namely soft contact technology. Based on ADAMS dynamics software, this paper compares and simulates the mechanism model of on-orbit soft-contact mechanism based on the bionic model and the integrated double joint model. The main purpose is to verify the path planning ability and the momentum buffering ability based on the different design concept mechanism. The simulation results show that both the two mechanism models have the path planning function before the space target contact, and also has the momentum buffer and controllability during the space target contact process.

1. Introduction

With the development and maturity of space technology, the on-orbit servicing and guarantee is becoming an important means to save cost and restore losses. Its main mission is to enhance the control and support capability of space environment by finishing the on-orbit maintenance, guarantee and service task through enhancing the space environment control and support capability through a variety of space operations. The space contact technology is the basic aspect of the on-orbit servicing and guarantee. [1,2] Currently, the world's major space powers have already realized the space contact technology on-orbit tests, such as NASA's "Orbital Express" [3] and Japan's ETS-VII [4]. But these tests basically adopt "hard-contact technology", that is, directly contacting with rigid components in orbit capture technology. This technology's space target operating condition is very harsh. The DART launched by the US in 2005 is a classic example of the dangers of hard-contact technology. This satellite started to rendezvous and dock with a communications satellite after entering into the orbit. However, during this process, the DART broke down and collided with the target satellite. Finally, this mission was failed because the impulsive momentum generated by the collision made the orbit height of the target satellite increasing about 3 km. [5] The main reason is that the main spacecraft's space contacting device is generally composed of a single rigid arm or multi-rigid arm cascade, and the space target relative attitude requires keeping highly stability during operating. This requires that any abnormal disturbance should not happen when the main spacecraft and space targets are contacting with each other. This greatly limits the space contact technology to improve on-orbit operational capacity and the range of applications of space target.

To resolve the on-orbit hard-contact technology problem, it is necessary to find a new technical method controlling and buffering the collision momentum generated during the collision moment.

This paper proposes a new concept, which is called soft-contact technology. Its core idea is to build up the soft-contact mechanism model and rigid-flexible complex control methods to achieve transformation from stepped collision momentum to harmonic attenuation momentum, and the key parameters, such as harmonic amplitude, phase and convergence time, can be controlled according to the desired function. [6] This idea can make collision momentum controlled buffering and unloading come true. According to the literature, it is found that robots based on modular joints and bionics are excellent in rigid-flexible compound control. For example, the movement of the German Powercube modular robots is very simple, the flexibility of its modular joints is for the path planning, refers to the flexibility of the harmonics reducer of the flexible wheel. [7] Fujita developed a reconfigurable robot platform, which is based on the SONY company OPEN-R standard to build a variety of software and hardware modules, and through these modules composed a variety of different robot structure. [8] Shigeo Hirose of Tokyo Institute of Technology in Japan completed the world's first bionic arm structure in 1972 and proposed a "serpentine curve" to simulate the movement of the biological snake. [9] In 2000, the German National Laboratory developed a flexible snake robot GMD, which can be bent in the vertical and horizontal direction, mainly for underwater detection operations. [10]

As the flexible robot of modular joint and snake-like robot have great flexibility and controllability, the paper proposed two design schemes of space manipulator based on bionics and based on integrated double joint to explore solutions of controlled movement limitation issues of the current space manipulator. Firstly, the joint configuration of space manipulator based on bionics is designed. Secondly, the kinematic equation of the manipulator is built up. Finally, the manipulator rigid and flexible characteristics are simulated in order to assess whether the soft-contact technology could meet the design idea under the space environment or not.

2. design of Joint conceptual model

On the one hand, joint conceptual model design should meet the requirement of path planning in dexterity, work space and complexity. On the other hand, it should have the ability to reduce disturbed momentum of the satellite platform in buffering control process, and transforms the pulsed pass to harmonic pass to meet the need of on-orbit mission through synergistic control. Therefore, the joint conceptual model must contain two functional modules: a rigid drive unit and a flexible damping unit. This model is built by ADAMS in three-dimension. The space soft-contact mechanism general design is shown as Figure 1.

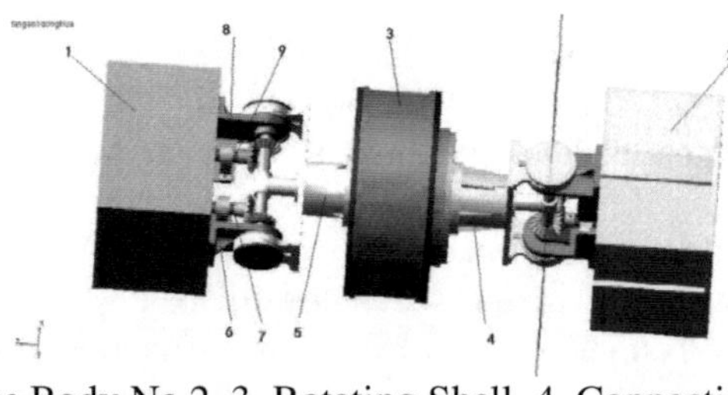

1.Base Body No.1, 2. Base Body No.2, 3. Rotating Shell, 4. Connecting rod No.2, 5. Connecting rod No.1, 6. Bracket No.1, 7. Buffer Shaft, 8. Bracket No.2, 9. Drive Shaft

Figure 1. Model of the space soft-contact mechanism

The rigid transmission unit uses motor as the drive. In consideration of the complexity of the space environment, the deceleration mechanism uses planetary gear. Rotation between base body and cross shaft is driven by mutual meshing motion between the gears, so as to realize the rotation between two manipulators and to complete pitch (X direction) or the yaw (Y direction) movement. The construction of the flexible controllable damping unit needs to follow the three principles below. Firstly, the flexible controllable damping unit is built on the base of the rigid transmission unit. So it could not destroy the function of the rigid transmission unit, while must keeping the flexible characteristics. Secondly, in order to buffer and unload the disturbance momentum received by the satellite platform during the on-orbit contact, the flexible controllable damping unit should possess the momentum unloading capacity of six degrees of freedom. Thirdly, the damping force generated by the flexible controllable damping unit should be controlled damping force, for the flexible damping unit can transfer the impulsive disturbance momentum into the harmonic disturbance momentum after the satellite platform being touched by the target. So that the disturbance momentum of the satellite platform can be reduced.

The space soft-contact mechanism based on the integrated double joints general design is shown as Figure 2.

1. motion block 1, 2. static block 1, 3. static block 2, 4. motion block 2, 5. Connecting rod, 6. Rotating Shell

Figure 2. Model of the space soft-contact mechanism

The drive transmission mechanism of rigid drive unit uses motor as the drive and the harmonic reducer is used as reduction mechanism. When the motor works, it drives the rotor drive shaft, then makes the wave generator start to rotate for reduction, while outputting by the rigid wheel and driving torque sensors and motion block rotation, to complete the pitch direction (X direction) or the yaw direction (Y direction) movement. Flexible damping unit is built up according to the following three strategies. Firstly, momentum of axial direction can be buffered by constructing a flexible controlled linear damper of axial direction. The substrate is divided into two parts. These two parts can slide relatively by nested with each other, so that the momentum can be applied to the damper and buffer. Secondly, the rotating flexible controllable dampers of the pitch and yaw direction should be constructed. When the joint is collided by the target satellite from pitch (and yaw) direction of rotation, the angular motion transfers directly to the spindle, at this moment the collision torque can be transmitted to the flexible linear controllable damper, by using bevel gears and pinion mechanism. Thirdly, the rotating flexible controllable damper of axial direction should be constructed. When the joint is collided by the revolving target satellite from axis direction, the angular motion transfers directly to the substrate. The rotation momentum of axis direction can be buffered by the rotating flexible controllable damper between substrates. [6]

3. The establishment of the kinematic equation

Kinematic parameters are shown in Figure 3 [11]. The study of this paper is carried out in the free-floating mode.

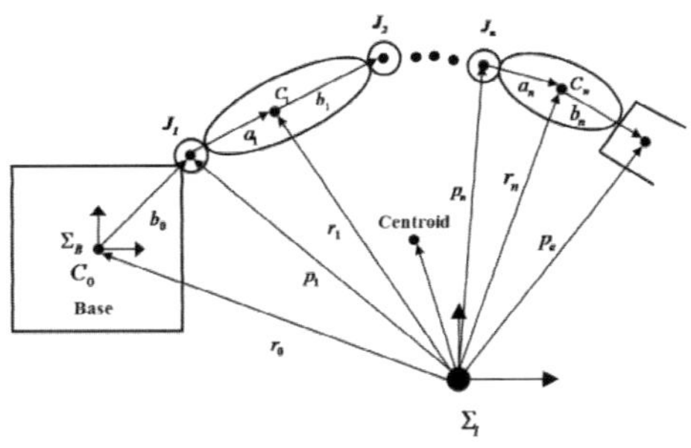

Figure 3. Manipulator kinematic parameters

In Figure 3, Σ_I, Σ_B, Σ_i represent the inertial coordinate system, the coordinate system of the base, and manipulator coordinate system, respectively. $\boldsymbol{a}_i \in \boldsymbol{R}^3$ represents the position vector from the joint i to the manipulator i's centroid. $\boldsymbol{b}_0, \boldsymbol{b}_i \in \boldsymbol{R}^3$ respectively represent the position vector from base centroid to joint 1 and the position vector from the arm centroid to joint. And then $\boldsymbol{r}_0, \boldsymbol{r}_i, \boldsymbol{r}_g \in \boldsymbol{R}^3$ respectively represent the position vector of the base centroid, the position vector of manipulator i's centroid, and the position vector of multibody system's centroid. $\boldsymbol{p}_i, \boldsymbol{p}_e \in \boldsymbol{R}^3$ respectively represents the position vector of joint i and the position vector the actuator end of manipulator.

And in order to the kinematic equation of this manipulator, this paper defines some parameters, such as $\boldsymbol{E}_i \in \boldsymbol{R}^{3\times3}$ represents a unit matrix. m_i represents manipulator i's quality. M represents the total mass of the multibody system. $\boldsymbol{Z}_i \in \boldsymbol{R}^3$ represents the joint i's rotation axis unit vector. $\boldsymbol{I}_i \in \boldsymbol{R}^{3\times3}$ represents the inertia of the manipulator relative to its centroid. $\boldsymbol{\Theta} \in \boldsymbol{R}^n$ represents the joint angle vector in the joint space of manipulator. $\boldsymbol{v}_0, \boldsymbol{\omega}_0 \in \boldsymbol{R}^3$ represent linear velocity and angular velocity of multibody system of base. $\boldsymbol{v}_e, \boldsymbol{\omega}_e \in \boldsymbol{R}^3$ represent linear velocity and angular velocity of the actuator end of manipulator. According to the kinematics parameters of manipulator in Figure. 3 and the space manipulator energy conservation law under free-floating mode, the generalized Jacobian matrix between the base and each joint angular velocity is established as below. [7]

$$\begin{bmatrix} \boldsymbol{v}_0 \\ \omega_0 \end{bmatrix} = \begin{bmatrix} -\tilde{\boldsymbol{r}}_{0g}\boldsymbol{H}_s^{-1}\boldsymbol{H}_\Theta - \dfrac{\boldsymbol{J}_{Tw}}{M} \\ -\boldsymbol{H}_s^{-1}\boldsymbol{H}_\Theta \end{bmatrix} \dot{\boldsymbol{\Theta}} = \boldsymbol{J}_{bm}\dot{\boldsymbol{\Theta}} \tag{1}$$

$\boldsymbol{J}_{bm}$ is the base-manipulator Jacobian matrix, and disturbances of the base generated by the movement of the manipulator can be obtained by this matrix. and then

$$\begin{cases} H_s = \left(M\tilde{\boldsymbol{r}}_{0g}\tilde{\boldsymbol{r}}_{0g} + \boldsymbol{H}_w\right) \\ \boldsymbol{H}_\Theta = \left(\boldsymbol{H}_{w\phi} - \tilde{\boldsymbol{r}}_{0g}\boldsymbol{J}_{Tw}\right) \end{cases} \tag{2}$$

$\boldsymbol{H}_w$ and $\boldsymbol{H}_{w\phi}$ in (2) are as below.

$$\boldsymbol{H}_w = \sum_{i=1}^{n}\left(\boldsymbol{I}_i + m_i\tilde{\boldsymbol{r}}_{0i}^{\mathrm{T}}\tilde{\boldsymbol{r}}_{0i}\right) + \boldsymbol{I}_0,\ \ \boldsymbol{H}_{w\phi} = \sum_{i=1}^{n}\left(\boldsymbol{I}_i\boldsymbol{J}_{Ri} + m_i\tilde{\boldsymbol{r}}_{0i}\boldsymbol{J}_{Ti}\right) \tag{3}$$

And also $\boldsymbol{J}_{Ri}$ and $\boldsymbol{J}_{Ti}$ in (3) are as below.

$$\begin{cases} \boldsymbol{J}_{Ri} = \left[\boldsymbol{Z}_1, \boldsymbol{Z}_1, \cdots, \boldsymbol{Z}_i, 0, \cdots, 0\right] \\ \boldsymbol{J}_{Ti} = \left[\boldsymbol{Z}_1 \times (\boldsymbol{r}_i - \boldsymbol{p}_1), \boldsymbol{Z}_2 \times (\boldsymbol{r}_i - \boldsymbol{p}_2), \ldots, \boldsymbol{Z}_i \times (\boldsymbol{r}_i - \boldsymbol{p}_i), 0, \ldots, 0\right] \end{cases} \tag{4}$$

At the same time, the generalized Jacobian matrix between velocity of the actuator end of manipulator and the angular velocity of each joint is built up as (5).

$$\begin{bmatrix} \boldsymbol{v}_e \\ \omega_e \end{bmatrix} = [\boldsymbol{J}_m + \boldsymbol{J}_b\boldsymbol{J}_{bm}]\dot{\boldsymbol{\Theta}} = \boldsymbol{J}_g\dot{\boldsymbol{\Theta}} \tag{5}$$

Where $\boldsymbol{J}_g$ is the generalized Jacobian matrix which descripts the relationship between velocity of the actuator end of manipulator and the angular velocity of joints. And then

$$\boldsymbol{J}_m = \begin{bmatrix} \left[\boldsymbol{Z}_1 \times (\boldsymbol{p}_e - \boldsymbol{p}_1), \boldsymbol{Z}_2 \times (\boldsymbol{p}_e - \boldsymbol{p}_2), \ldots, \boldsymbol{Z}_n \times (\boldsymbol{p}_e - \boldsymbol{p}_n)\right] \\ \left[\boldsymbol{Z}_1, \boldsymbol{Z}_2, \ldots, \boldsymbol{Z}_n\right] \end{bmatrix},\ \ \boldsymbol{J}_b = \begin{bmatrix} \boldsymbol{E}_3 & -(\tilde{\boldsymbol{p}}_{0e}) \\ 0 & \boldsymbol{E}_3 \end{bmatrix} \tag{6}$$

4. Rigid-flexible characteristic simulation comparison of single-joint manipulator

4.1. Simulation of Single Joint Rigid Drive Unit

As to the space soft-contact mechanism based on the bionics, Soft-contact mechanism should have the ability of path planning like the traditional multi-rigid manipulator, in order to achieve its large-scale movement. So its single joint must have rigid transmission function. By adding the motor drive, while combining function and script control, joint range of motion can be simulated. Yaw motor drive function of the simulation is "if(time-10:9d,0, if(time-20:-9d,0,if(time-30:9d,0,0)))", and the step is 300. The status of simulation of single joint space soft contact mechanism is shown as Figure 4. Figure 4 shows that, in the ADAMS's simulation, the single joint's degree of freedom can range up to -90° ~ + 90°. As to the space soft-contact mechanism based on the integrated double joint, it should have the ability of path planning like the traditional multi-rigid manipulator, in order to achieve its large-scale movement. So its single joint must have rigid transmission function. According to the structure and transmission principle, all the damping buffer mechanisms lose efficacy when the manipulator start moving, and the electromagnetic brake and the positioning mechanism are activated. The system is in the rigid structure state. By adding the motor drive, while combining function and script control, joint range of motion can be simulated. Yaw motor drive function of the simulation is "if(time-10:9d,0,if(time-20:-9d,0,if(time-30:9d,0,0)))", and the step is 300.The simulation result is shown as Figure 5. Figure 5 shows that, in the ADAMS simulation, the single joint's degree of freedom can range up to 0° ~ 360°.

The simulation commonality of the two kinds of joint mechanism model of the drive structure is adding the drive to the motor shaft, and the drive function type is the same. But the values of the drive according to the design of the scope of rotation are a slight different. The simulation results of two kinds of joint mechanism model show that the rigid motion design structure can meet the requirement of large range of motion characteristics of soft-contact mechanism.

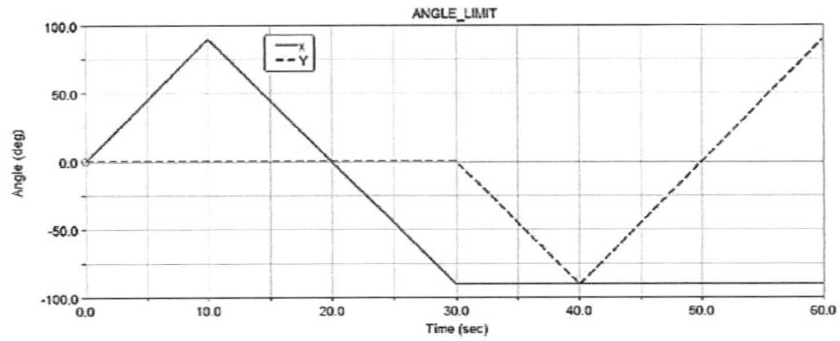

Figure 4. Rotation angle change of rigid simulation of the joint around the X and Y axis

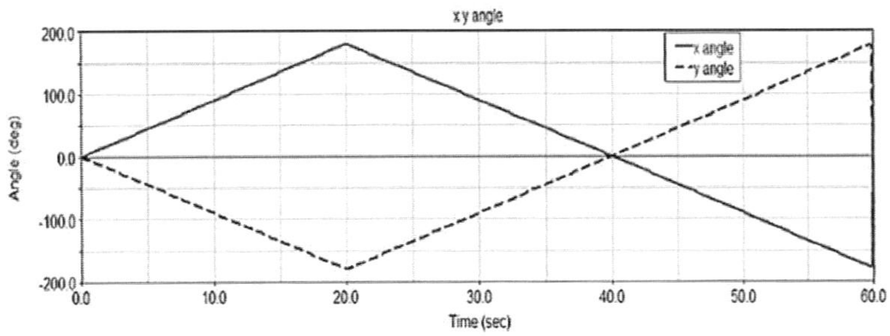

Figure 5. Rotation angle change of rigid simulation of the joint around the X and Y axis

4.2. Simulation of single joint flexible controllable unit

For the specific momentum collision at the end of a single joint, the simulation can be verified according to the application background of the following three typical space targets: ① Adding the instantaneous axial force to the end of Z axis to verify the working mechanism of the linear motion flexible controllable unit; ②Adding the instantaneous axial force and instantaneous torque to the end of Z-axis to verify the working mechanism of angular motion flexible control unit; ③ Adding instantaneous force and instantaneous torque with 45° to the end in each X / Y / Z axis to verify the working mechanism of the Z-axis linear motion flexible control unit and X / Y / Z axes three angular motion flexible control units. Because of the limit of paper length, we only present the simulation of compound flexible controllable unit. That is the third application background above. The results of the other two simulations are the same as the third one.

We carry out simulation verification on the third application background of space target control, that is, the working mechanism of the angular motion flexible control unit. Then adding instantaneous force and torque to the end of 45° to the X/Y/Z axes to simulate the Z-axis Linear motion and X/Y/Z axes angular motion composite flexible control mechanism. The initial conditions are shown in Table 1.

Table 1. Composite flexible controllable unit initial condition

instantaneous force	Driving function: if(time-0.1:20N,0,0), 45° to the X/Y/Z axis

IOP Conf. Series: Journal of Physics: Conf. Series **1016** (2018) 012015 doi:10.1088/1742-6596/1016/1/012015

instantaneous torque	Driving function: if(time-0.1:0.5N-m,0,0), 45° to the X/Y/Z axis
linear motion buffer damping coefficient	Stiffness coefficient: 0.5N/mm, Damping coefficient: 0.05N-s/mm
Roll direction angular motion buffer damping coefficient	Stiffness coefficient: 3 N/mm, Damping coefficient: 2.5 N-s/mm
Pitch direction angular motion buffer damping coefficient	Stiffness coefficient: 3 N/mm, Damping coefficient: 2.5 N-s/mm
Yaw direction angular motion buffer damping coefficient	Stiffness coefficient: 8 N/mm, Damping coefficient: 5 N-s/mm

According to the initial conditions given in Table 4, the simulation results of the composite flexible control of the on-orbit soft contact mechanism based on different mechanism models can be obtained by simulating the collision process of space targets with both linear motion and angular motion. The results are shown in Figure 6.

The dynamic parameters of the two mechanism models in flexible controllable mode were compared and analysed in detail. The simulation results show that both the two mechanism models can achieve the expected goal, but the mechanism model based on bionics is slightly better in the comprehensive performance.

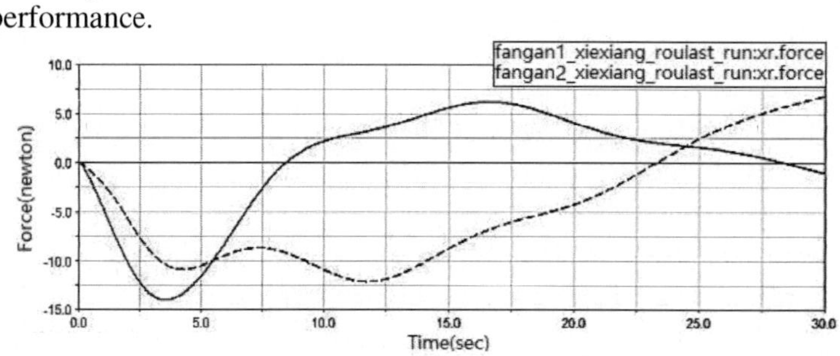

(a) angular motion buffer changes of Roll direction

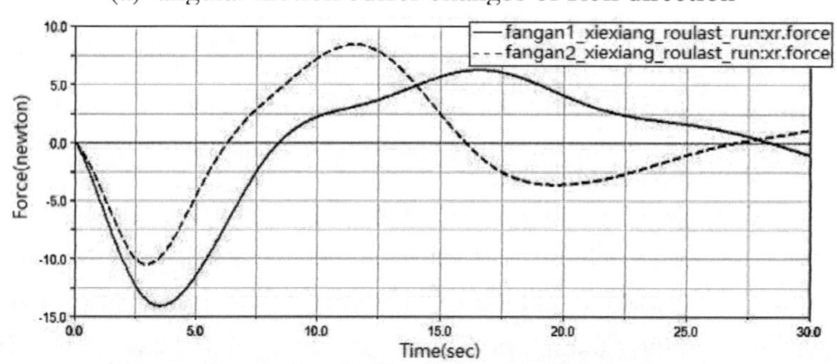

(b) angular motion buffer changes of Pitch direction

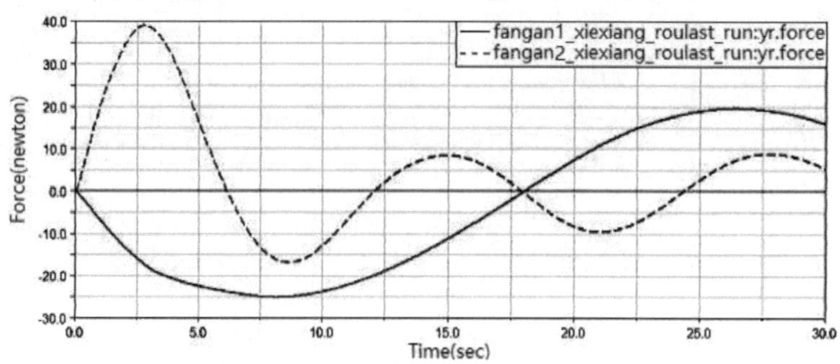

(c) angular motion buffer changes of Yaw direction

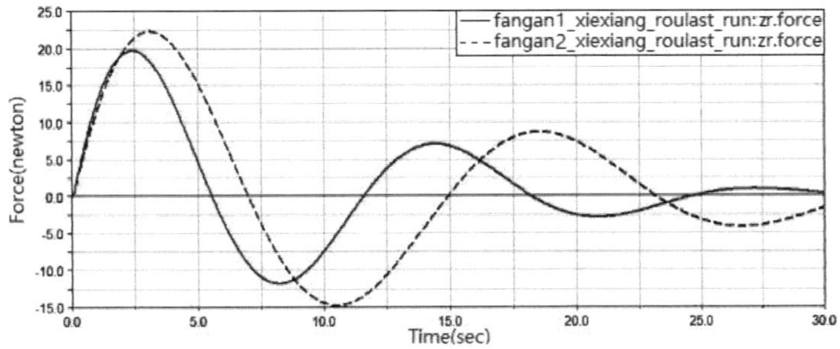

(d) linear motion buffer changes of Yaw direction

Figure 6. Influence of Dynamic Parameters of Satellite Base of two Mechanism Models

5. conclusion

In this paper, the dynamic characteristics of two different single joint soft-contact mechanisms are simulated. On the one hand, the simulation results and evaluation analysis of the rigid transmission unit in the path planning of the soft-contact mechanism are given based on both the bionic model and the integrated double joint model. On the other hand, the mechanism model simulation of the single joint flexible controllable unit is carried out, and the comparison results of the control dynamics under the three typical space target control applications are given. The simulation results show that both the two mechanism models have the path planning function before the space target contact, and also has the momentum buffering and controllability during the space target contact process. However, the bionic model is slightly better in the comprehensive performance.

References

[1] Richard W M 2000 Micro-satellite based, On-orbit servicing work at the air force research *Aerospace Conference Proceedings* (Big Sky: IEEE) vol 4 pp 215-26

[2] Naigang C, Ping W, Jifeng G and Xing C 2007 A Review of On-Orbit Servicing *Journal of Astronautics* **28** 33-39

[3] Stamm S, Pejmun M 2004 Orbital Express Capture System: concept to reality *In Spacecraft Platforms and Infrastructure Proc. of SPIE* (Bellingham: SPIE) vol 5419 pp 78-91

[4] Oda M 1999 Space robot experiments on NASDA's ETS-VII satellite-preliminary overview of the experiment results *In Proc. Of IEEE International Conf. on Robotics and Automation* (Detroit: IEEE) vol 2, 1390-1395.

[5] Croomes S 2006 Overview of the DART mishap investigation results *NASA Report* pp 1-10.

[6] Xin Y, Zhenghong D, Jiacai H and Fan Y 2016 Research on kinematic characteristic of a new flexible space manipulator based on the integrated double joints *In Proc.of 2016 5th International Conf. on Mechatronics and Control Engineering* (Venice: ACM) 86-91

[7] Zhiqiang Z 2008 *Research on PowerCube Reconfigurable Modular Robots* (Chengdu: Xihua University Press)

[8] Fujita M, Hiroaki K, and Koji K 1999 A reconfigurable robot platform *Robotics and Autonomous Systems* **29** 119-32.

[9] Shigeo H and Yoji U 1978 The development of soft gripper for the versatile robot hand *Mechanism and Machine Theory* **13** 351-59

[10] K. L. Paap, M. Dehlwisch and B. Klaassen 1996 GMD-Snake： A Semi-Autonomous Snake-like Robot *3rd Int. Symp. on Distributed Autonomous Robot Systems* (Tokyo: Springer) vol 2 pp 29-31

[11] Wenfu X 2007 *Path Planning and Experiment Study of Space Robot for Target Capturing* (Harbin: Harbin Institute of Technology Press)

A method of rotating accelerometer gravity gradiometer for centrifugal gradient detection

Mingbiao Yu[1], Tijing Cai[1,2]

[1]School of Instrument Science and Engineering, Southeast University, Nanjing, 210096, China;

[2]Correspondence: caitij@seu.edu.cn; Tel.: +86-025-83794458.

Email: ymb_moon@126.com;

Abstract. This paper proposed a centrifugal gradient detection method of rotating accelerometer gravity gradiometer (RAGG) and deduced the relationship between centrifugal gradient detection precision and gyro noise coefficients. Based on the deduced relationship, suitable gyros can be selected as angular velocity sensors for centrifugal gradient detection. A simulation experiment was designed to verify the centrifugal gradient detection method and the deduced relationship. In the experiment, the permitted centrifugal gradient error was given, the gyro critical noise coefficients were calculated, and their corresponding gyro simulation data was generated; based on the proposed centrifugal gradient detection method, centrifugal gradient and centrifugal gradient detection error were calculated. The experiment results show that the centrifugal gradient and the centrifugal gradient error are respectively consistent with the theoretical centrifugal gradient and the given permitted centrifugal gradient error; therefore, the results suggest that the centrifugal gradient detection method and the deduced relationship is correct.

1. Introduction

Gravity gradiometry began in 1880s when Loránd Eötvös introduced his first torsion balance gradiometer to the oil industry [1]. Subsequently, in the 1970s, the United States initiated a program of developing moving-base gravity gradient instrument for military needs, and by the 1980s, Bell Aerospace Textron had successfully developed the first moving base rotating accelerometer gravity gradiometer [1, 2]. Compared with gravity data, gravity gradient data provides more precise mapping, more accurate depth information, more accurate shape information and more accurate orientation information. Airborne gravity gradiometry (AGG) has become the most advanced means of surveying gravity field and it has high accuracy and space resolution for measuring the gravity field. Gravity precision derived from Falcon AGG (a partial tensor RAGG), for heliborne surveys, is reported to be 0.1 *mGal* with spatial resolution 50*m* [2]. A more detailed review on current technologies can be found in [2-7].

The measurement of RAGG contains centrifugal gradient, to realize the measurement of the gravitational gradient, the output of RAGG needs to subtract the centrifugal gradient measured by centrifugal gradient detection unit. Patent [8] proposed applying gyro as angular rate sensor for centrifugal gradient detection. Furtherly, this paper studied centrifugal gradient detection method of RAGG, and deduced the relationship between centrifugal gradient detection precision and gyro noise

IOP Conf. Series: Journal of Physics: Conf. Series **1016** (2018) 012016 doi:10.1088/1742-6596/1016/1/012016

coefficients. Based on this relationship, suitable gyros can be selected as angular velocity sensors for centrifugal gradient detection.

2. Centrifugal gradient detection method

The measurement of RAGG is the difference of the sum of two pairs of accelerometers:

$$(A_1 + A_2) - (A_3 + A_4) = 2R\left(\omega_{imy}^2 - \omega_{imx}^2 + \Gamma_{xx} - \Gamma_{yy}\right)\sin 2\Omega t - 4R\left(-\omega_{imx}\omega_{imy} + \Gamma_{xy}\right)\cos 2\Omega t \tag{1}$$

where ω_{imx}, ω_{imy} respectively represent the angular velocity of RAGG relative to the inertial frame, $\omega_{imy}^2 - \omega_{imx}^2$, $-\omega_{imx}\omega_{imy}$ are centrifugal gradients, Γ_{xx}-Γ_{yy}, Γ_{xy} are gravitational gradients, RAGG's inline output ω_{imy}^2-ω_{imx}^2+Γ_{xx}-Γ_{yy} and RAGG's cross output $-\omega_{imx}\omega_{imy}+\Gamma_{xy}$ respectively are the demodulation of the combination signal (Equation 1) at $sin2\Omega t$ and $cos2\Omega t$. Therefore, the output of RAGG is the sum of the centrifugal gradient and the gravitational gradient. To realize the measurement of the gravitational gradient, the output of RAGG needs to subtract the centrifugal gradient. Angular velocity sensors (gyros) are mounted on RAGG for measuring centrifugal gradient. Figure 1(a) illustrates the installation location of gyros. X_m Y_m Z_m is RAGG measurement frame, two gyros are mounted such that its input axis is orthogonal to each other for measuring ω_{imx} and ω_{imy} [1]. Figure 1(b) illustrates the proposed signal processing method of centrifugal gradient detection.

Comparing to the outputs of RAGG, the raw data of gyros has a higher sampling frequency, and contains high-frequency noise such as quantify noise, power frequency noise etc., which has a significant impact on centrifugal gradient detection precision. Lowpass filter 1 is applied to attenuate these high-frequency noise, then anti-aliasing filtering and decimation are performed to make sample rate of gyros data equalling that of RAGG's output. The calculation of centrifugal gradient is given by

$$\begin{cases} \Gamma_{inline}^c = k_{wc} \cdot \left(\omega_{imx}^2 - \omega_{imy}^2\right) \\ \Gamma_{cross}^c = k_{wc} \cdot \left(-\omega_{imx}\omega_{imy}\right) \end{cases} \tag{2}$$

where the unit of angular velocity is *arsec/s*, k_{wc} is conversion factors from $(arsec/s)^2$ to *Eo*, and its value equals 0.0235. The centrifugal gradient that is calculated from the decimated gyro data, subsequently is lowpass filtered by Lowpass filter 2, and the filter characteristic of lowpass filter 2 should be in accordance with that of RAGG to ensure that the RAGG and gyro experience the same centrifugal gradient.

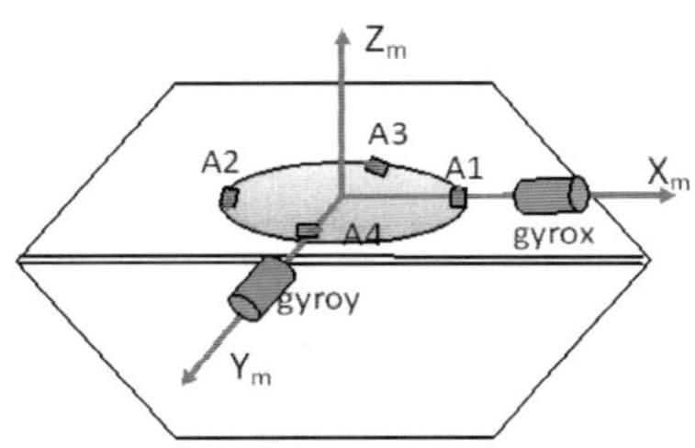

(a)

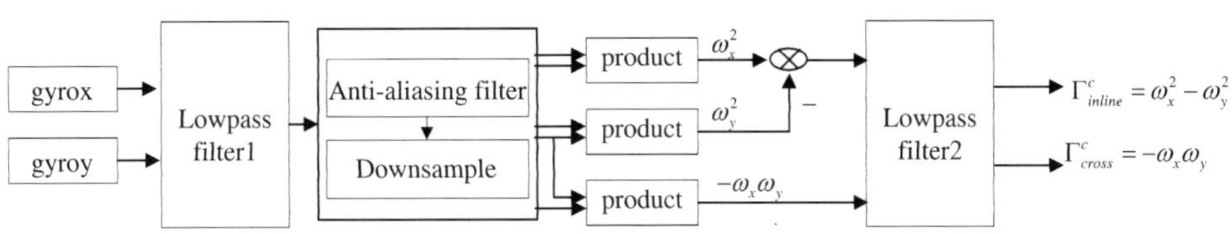

(b)

Figure 1. Centrifugal gradient detection method

3. Relationship between centrifugal gradient detection precision and gyro noise coefficients

In this section, we will analyse the relationship between centrifugal gradient detection accuracy and the gyro noise coefficients. The output of gyro is given by:

$$\begin{cases} \omega_x = \omega_{rx} + \delta w_x \\ \omega_y = \omega_{ry} + \delta w_y \end{cases} \tag{3}$$

where ω_r is true value, $\delta\omega$ is noise value, by equation (2), centrifugal gradient can be calculated by

$$\begin{cases} \begin{cases} \Gamma_{inline} = k_{wc}\left(\omega_x^2 - \omega_y^2\right) \\ \Gamma_{cross} = -k_{wc}\left(\omega_x \omega_y\right) \end{cases} \\ \begin{cases} \Gamma_{inline}^r = k_{wc}\left(\omega_{rx}^2 - \omega_{ry}^2\right) \\ \Gamma_{cross}^r = -k_{wc}\left(\omega_{rx}\omega_{ry}\right) \end{cases} \end{cases} \tag{4}$$

where Γ_{inline} and Γ_{cross} are centrifugal gradients calculated from the output of gyros, Γ_{inline}^r and Γ_{cross}^r are true value of centrifugal gradients, the centrifugal gradient error is the difference of those two kinds centrifugal gradients, and is given by

$$\begin{cases} err_{inline} = \Gamma_{inline} - \Gamma_{inline}^r = k_{wc}\left[\delta w_y^2 - \delta w_x^2 + 2\left(\omega_{ry}\delta w_y - \omega_{rx}\delta w_x\right)\right] \\ err_{cross} = \Gamma_{cross} - \Gamma_{cross}^r = k_{wc}\left[\delta\omega_x \delta\omega_y + \omega_{rx}\delta\omega_y + \omega_{ry}\delta\omega_x\right] \end{cases} \tag{5}$$

the standard deviation of centrifugal gradient error is given by

$$\begin{cases} \sigma\left(err_{inline}\right) = k_{wc}\left[\sigma\left(\delta w_y^2 - \delta w_x^2\right) + 2\omega_{ry}\sigma\left(\delta w_y\right) + 2\omega_{rx}\sigma\left(\delta w_x\right)\right] \\ \sigma\left(err_{cross}\right) = k_{wc}\left[\sigma\left(\delta\omega_x \delta\omega_y\right) + \omega_{rx}\sigma\left(\delta\omega_y\right) + \omega_{ry}\sigma\left(\delta\omega_x\right)\right] \end{cases} \tag{6}$$

Assuming that the x-gyro and y-gyro have about the same angular velocity $\omega_{rx}=\omega_{ry}=\Omega$, and their noise are independent zero mean gaussian white noise, and their standard deviation satisfy with $\delta(\delta\omega_x)=\delta(\delta\omega_y)=\sigma_{gyro}$. Thus, the equation (6) can be simplified as

$$\begin{cases} \sigma\left(err_{inline}\right) = \sqrt{2}k_{wc}\sigma\left(\delta w_y^2\right) + 4k_{wc}\Omega\sigma\left(\delta w_y\right) \\ \sigma\left(err_{cross}\right) = k_{wc}\sigma\left(\delta w_y^2\right) + 2k_{wc}\Omega\sigma\left(\delta w_y\right) \end{cases} \tag{7}$$

Base on the property of random variable, equation (7) can be simplified as:

$$\begin{cases} \sigma\left(err_{inline}\right) = 2k_{wc}\sigma_{gyro}^2 + 4k_{wc}\Omega\sigma_{gyro} \\ \sigma\left(err_{cross}\right) = \sqrt{2}k_{wc}\sigma_{gyro}^2 + 2k_{wc}\Omega\sigma_{gyro} \end{cases} \tag{8}$$

Let Γ_{max} denotes the maximum permitted standard deviation of centrifugal gradient error, therefore the standard deviation of centrifugal gradient error should satisfy below condition

$$\begin{cases} \sigma\left(err_{inline}\right) \le \Gamma_{\max} \\ \sigma\left(err_{cross}\right) \le \Gamma_{\max} \end{cases} \tag{9}$$

Because of $4\Omega\sigma_{gyro}>>2\sigma_{gyro}^2$, by equation (9), we can get

$$\sigma_{gyro} \le \Gamma_{\max}/4\Omega k_{wc} \tag{10}$$

Equation (10) reflects the relationship between maximum permitted standard deviation of the centrifugal gradient error, gyro angular velocity and standard deviation of gyro noise. The standard deviation of gyro noise, σ_{gyro} is the superposition of different noise, such as angle random walk, rate random walk, sine noise etc. The relationship between maximum permitted standard deviation and noise coefficients of four basic noise terms are discussed in below. Four basic noise terms are quantization noise, angle random walk, rate random walk, and bias instability. Four basic noise terms are discussed below.

Quantization Noise: quantization noise is one of the errors introduced into an analogy signal by encoding it in digital form[9]. The variance of the quantization noise is given by

$$\sigma_q^2 = \frac{8\pi^2 T_s Q^2 f_q^3}{3} \tag{11}$$

IOP Conf. Series: Journal of Physics: Conf. Series **1016** (2018) 012016 doi:10.1088/1742-6596/1016/1/012016

where Q is quantization-noise coefficient, T_s is sample interval, f_q is equivalent bandwidth of quantization-noise, standard deviation of quantization-noise should satisfy condition of equation (10), so we can get

$$Q \le \left(\frac{3\Gamma_{\max}^2 f_s}{128\pi^2 k_{wc}^2 \Omega^2 f_q^3} \right)^{1/2} \tag{12}$$

Angle Random Walk: angle random walk is the integral of rate white noise, and its variance is given by

$$\sigma_{arw}{}^2 = N^2 f_{arw} \tag{13}$$

where N is angle random walk noise coefficient, f_{arw} is equivalent bandwidth of white noise, standard deviation of angle random walk should satisfy the condition of equation (10), so we can get

$$N \le \left(\frac{\Gamma_{\max}^2}{16\Omega^2 k_{wc}^2 f_{arw}} \right)^{1/2} \tag{14}$$

Rate Random Walk: rate random walk is the integral of acceleration rate white noise[6], assuming that the gyro is running for T hour, the minimum frequency f_{min} of the acquired data is $1/T$, and its variance is given by

$$\sigma_{rr\omega}{}^2 = \frac{k^2}{2\pi^2} \frac{f_{rrw}}{f_{\min}\left(f_{rrw} + f_{\min}\right)} \tag{15}$$

where k is rate random walk noise coefficient, f_{rrw} is equivalent bandwidth of rate random walk noise, standard deviation of rate random walk noise should satisfy the condition of equation (10), so we can get

$$k \le \left(\frac{2\pi^2 \Gamma_{\max}^2 \left(f_{rrw} + f_{\min}\right)}{16\Omega^2 k_{wc}^2 f_{rrw}} \right)^{1/2} \tag{16}$$

Bias Instability: The origin of this noise is the electronics or other components that are susceptible to random flickering[9]. Assuming that the gyro is running for T hour, the minimum frequency f_{min} of the acquired data is $1/T$, and the variance of this noise is given by

$$\sigma^2_{BiasInst} = \frac{B^2}{\pi} \left(\ln \frac{\left(f_{BiasInst} + f_{\min}\right)}{f_{\min}} \right) \tag{17}$$

where B is bias instability noise coefficient, $f_{BiasInst}$ is equivalent bandwidth of bias instability, standard deviation of bias instability should satisfy the condition of equation (10), so we can get

$$B \le \left(\frac{\Gamma_{\max}^2 \pi}{16\Omega^2 k_{wc}^2 \left(\ln\left(f_{BiasInst} + f_{\min}\right) / f_{\min} \right)} \right)^{1/2} \tag{18}$$

Equation (12), (14), (16), and (18) are the relationship between four basic gyro noise coefficients and centrifugal gradient detection precision.

4. Simulation Experiments

The relationship between gyro noise coefficients and centrifugal gradient detection precision is derived in the above, validation experiment is performed in below. In validation experiment, the permitted centrifugal gradient error of single noise term is $0.1Eo$, the gyro noise is the superposition of four basic noise terms, the centrifugal gradient errors are the superposition of four basic noise terms, therefore, the maximum centrifugal gradient errors should approximate $0.2Eo$. Base on the deduced relationship, critical noise coefficients of gyro are calculated, and their corresponding gyro simulation data is

IOP Conf. Series: Journal of Physics: Conf. Series **1016** (2018) 012016 doi:10.1088/1742-6596/1016/1/012016

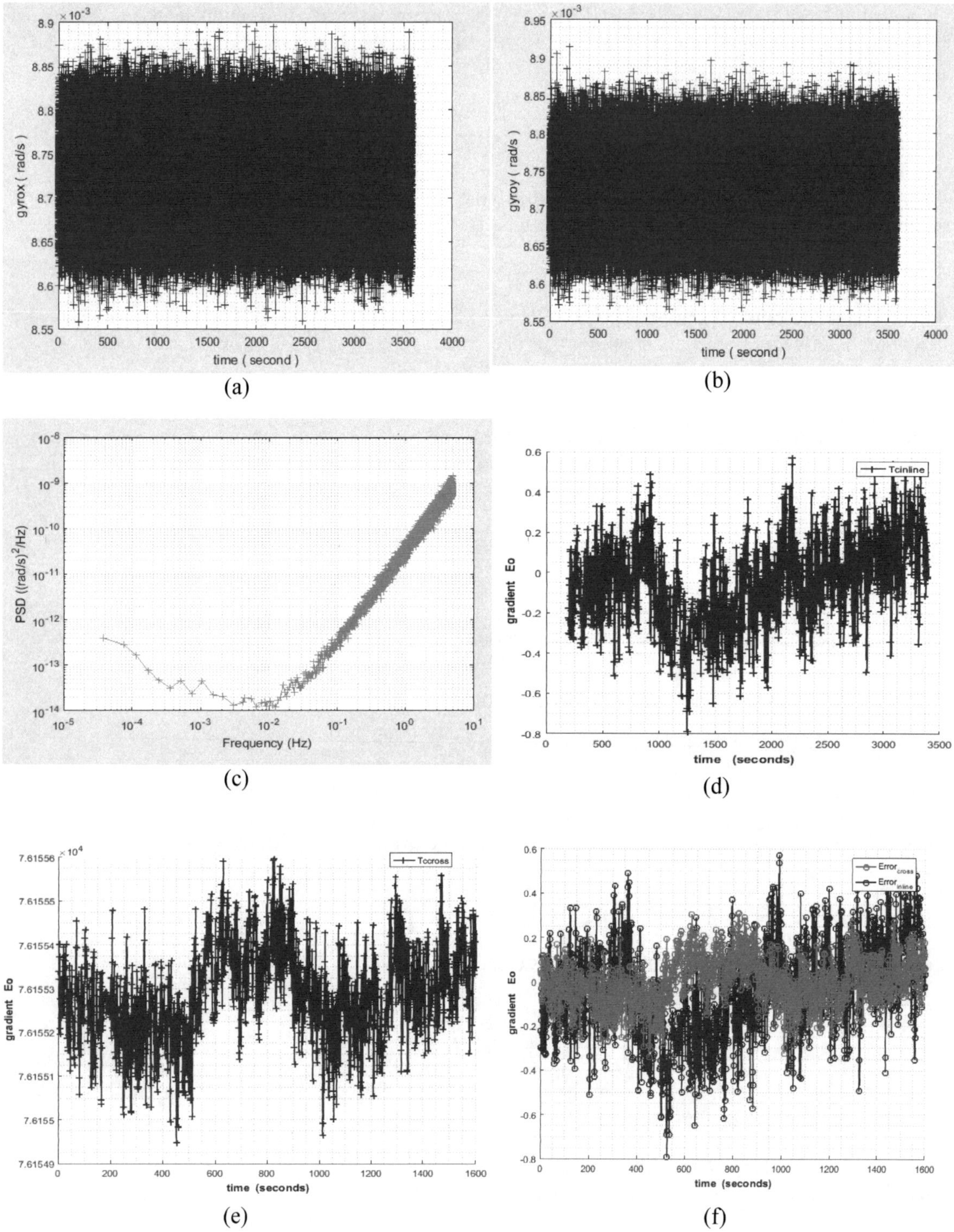

Figure 2. Validation experiment results

generated. If the centrifugal gradients calculated from the gyro simulation data are consistent with the theoretical centrifugal gradients, then proves that the proposed centrifugal gradient detection method is correct. If the maximum standard deviation of centrifugal gradient errors calculated from the gyro simulation data approximates $0.2Eo$, then proves that the deduced relationship between gyro noise coefficients and centrifugal gradient detection precision is correct.

The gyro simulation data is the combination of true angular velocity and gyro noise, the gyro noise is simulated by the power spectral density models, $\Phi(f)$, that provide reasonable characterizations of the gyro noise at different frequencies, f [10]. The gyro noise *PSD* is the superposition of four basic noise terms, and $\Phi(f)$ is given by:

$$\Phi(f) = 8\pi^2 T_s Q^2 f^2 + 2N^2 + \frac{k^2}{2\pi^2 f^2} + \frac{B^2}{\pi f} \tag{19}$$

A set of reasonable parameters is selected, $\omega_x=\omega_y=\Omega=0.5$ °/s, $\Gamma_{max}=0.1Eo$, $f_q=0.0891Hz$, $f_{arw}=0.1135Hz$, $f_{rrw}=0.4mHz$, $f_{BiasInst}=2.8mHz$, $T=1hour$, by the deduced relationship, the critical noise coefficients of gyro are calculated, $Q=0.0433$ *arsec*, $N=0.0018$ $arsec/\sqrt{s}$, $k=0.0045$ $(arsec/s)/\sqrt{h}$, $B=0.0007$ *arsec/s*. Substituting these critical noise coefficients into equation (19), then we can generate gyro simulation data. The method of calculating centrifugal gradient from gyro data is described in detail in section 2. Figure 2 (a) and figure 2(b) are respectively X-gyro simulation data and Y-gyro simulation data. Figure 2(c) is log-log scale single-sided PSD of simulated gyro. Figure 2(d) and figure (e) is centrifugal gradient plot calculated from gyro simulation data, the mean of centrifugal gradients respectively are $T^m_{inline}=0.0041$ *Eo* and $T^m_{cross}=7.615523\times10^4 Eo$, and the theoretical centrifugal gradients are $T_{inline}=0$ *Eo* and $T_{cross}=7.6155\times10^4 Eo$, the centrifugal gradients calculated from gyro data is consistent with the theoretical centrifugal gradients, this experiment result suggests that the proposed centrifugal gradient detection method is correct. The Figure 2(f) is centrifugal gradient error plot, and the maximum standard deviation of centrifugal gradient errors approximates $0.2Eo$, this experiment result suggests that the deduced relationship is correct. Moreover, another two experiments were done, one experiment had its angle random walk coefficient of critical noise coefficients increase 5 times, and another experiment had its bias instability coefficient of critical noise coefficients increase 5 times, and the gyro PSD characteristic curve and gyro simulation data curve are not shown, their standard deviation of centrifugal gradient errors are listed in table 1. The maximum standard deviation of centrifugal gradient errors are respectively $0.470Eo$ and $0.563Eo$. Since the noise bandwidth of angle random walk noise is wider than that of bias instability noise, the change of angle random walk noise coefficient has greater impact on the precision of centrifugal gradient detection than the change of bias instability noise coefficient.

Table 1. Standard deviation of centrifugal gradient errors and their corresponding gyro noise coefficients

Q (arsec)	**N ($arsec/s^{1/2}$)**	**K ($deg/h^{3/2}$)**	**B (arsec/s)**	**Inline's error SD[a] (Eo)**	**Cross's error SD[b] (Eo)**
0.0433	0.0018	0.0045	0.0007	0.215	0.107
0.0433	0.0018	0.0045	0.0035	0.470	0.263
0.0433	0.009	0.0045	0.007	0.563	0.268

[a] centrifugal gradient error standard deviation of RAGG's inline channel
[b] centrifugal gradient error standard deviation of RAGG's cross channel

5. conclusions

The measurement of RAGG contains centrifugal gradient, to realize the measurement of the gravitational gradient, centrifugal gradient detection and compensation is an unavoidable issue. This paper examines the centrifugal gradient detection method of rotating accelerometer gravity gradiometer (RAGG) and the relationship between centrifugal gradient detection precision and gyroscope noise coefficients. The validation experiments results suggest that the proposed centrifugal gradient detection method and the relationship between centrifugal gradient detection precision and gyro noise coefficients are right. This relationship is helpful for choosing suitable gyros as angular velocity sensors for centrifugal gradient detection.

IOP Conf. Series: Journal of Physics: Conf. Series **1016** (2018) 012016 doi:10.1088/1742-6596/1016/1/012016

References

[1] Rogers M M 2009 An Investigation Into the Feasibility of Using a Modern Gravity Gradient Instrument for Passive Aircraft Navigation and Terrain Avoidance
[2] Dransfield M H and Christensen A N 2013 Performance of airborne gravity gradiometers *Leading Edge* **32** 908-22
[3] Annecchione M, Moody M, Carroll K, Dickson D and Main B 2007 Benefits of a high performance airborne gravity gradiometer for resource exploration. In: *Proceedings of Exploration,* pp 889-93
[4] Difrancesco D 2007 Advances and challenges in the development and deployment of gravity gradiometer systems. In: *EGM 2007 international workshop,*
[5] Dransfield M 2007 Airborne gravity gradiometry in the search for mineral deposits.
[6] Difrancesco D, Grierson A, Dan K and Meyer T 2009 Gravity gradiometer systems – advances and challenges *Geophysical Prospecting* **57** 615–23
[7] DiFrancesco D, Meyer T, Christensen A and FitzGerald D 2009 Gravity gradiometry–today and tomorrow. In: *11th SAGA Biennial technical meeting and exhibition,*
[8] Brett J and Brewster J 2010 Accelerometer and rate sensor package for gravity gradiometer instruments. US)
[9] Hou H and El-Sheimy N 2003 Inertial sensors errors modeling using Allan variance
[10] Jekeli C 2006 Airborne Gradiometry Error Analysis *Surveys in Geophysics* **27** 257-75

IOP Conf. Series: Journal of Physics: Conf. Series **1016** (2018) 012017 doi:10.1088/1742-6596/1016/1/012017

Research on the Diesel Engine with Sliding Mode Variable Structure Theory

Zhexuan Ma, Xiaobing Mao, Le Cai

School of Energy and Power Engineering, Wuhan University of Technology, Wuhan 430063, China

E-mail address: maoxiaobing2009@163.com

Abstract. This study constructed the nonlinear mathematical model of the diesel engine high-pressure common rail (HPCR) system through two polynomial fitting which was treated as a kind of affine nonlinear system. Based on sliding-mode variable structure control (SMVSC) theory, a sliding-mode controller for affine nonlinear systems was designed for achieving the control of common rail pressure and the diesel engine's rotational speed. Finally, on the simulation platform of MATLAB, the designed nonlinear HPCR system was simulated. The simulation results demonstrated that sliding-mode variable structure control algorithm shows favourable control performances which are overcoming the shortcomings of traditional PID control in overshoot, parameter adjustment, system precision, adjustment time and ascending time.

1. Introduction

Electronically controlled diesel engine has undergone three major development stages [1]. The first-generation position-control diesel engine used electronic governor to replace the mechanical governor, but cannot change the injection characteristics of the cam injection system. The second-generation product is time-control diesel engine, which employed magnetic value for addressing the problems in fuel injection timing and duration. However, pilot injection and multi-injection can be difficult to realize because of injection pressure. The third-generation product, which is widely applied at present, is the common-rail diesel engine maintaining a high fuel injection pressure. This is also referred to as time/pressure-control diesel engine, which can overcome the limitations of cam outlines. The fuel injection pressure is irrelevant to rotational speed and load [2].

Common-rail pressure is generally controlled by closed-loop proportion-integration-differentiation (PID) controller [3]. Classical PID controllers show simple structures and need not construct the system's accurate mathematical model. However, it is fairly difficult to seek out the appropriate proportional band, integration time and differentiation time simultaneously so as to achieve the optimal control effects. Moreover, in some engineering applications such as simple temperature or pressure systems, once the parameters of PID is changed, the system's control effect can hardly be restored to factory settings.

Variable-structure control is an advance control algorithm based on accurate mathematical model. As a comprehensive method for nonlinear control systems, variable-structure control is practical and can acquire a series of favourable properties and qualities [4]. Sliding mode variable structure control (SMVSC) is a control strategy for variable-structure control systems, which is also a comprehensive method in modern control theory based on phase plane method. The basic idea is that the system's motion first reaches a certain switching surface, then, the motion follows a sliding mode and gradually approaches the origin, and finally, the control goal can be realized when the sliding motion is endowed with good character.

At present, the main control method of diesel engine running parameters is PID. This paper used Sliding mode variable structure control algorithm to study the nonlinear diesel engine.

2. Control of diesel engine

In a high-pressure common rail (HPCR) system without accumulator-type injector and boosting piston, the fuel oil pressure in public oil passage can be directly controlled to maintain a high level (over 100 MPa). Fuel injection quantity and injection timing are regulated via the electromagnetically-controlled three-way valve or two-way valve. Specifically, fuel injection quantity and injection timing change with the variation of backpressure, which is controlled by the three-way valve or two-way valve. The pressure of high-pressure oil rail is related to the circulating fuel supply quantity [5].

2.1. Establishment of the nonlinear diesel engine system model

The common-rail pressure is subjected to the effects of fuel supply quantity of the high-pressure fuel pump, fuel return flow and fuel injection quantity of the injector. Accordingly, for the convenience in the design of control law and the related stability analysis, the common-rail pressure model for a diesel engine can be simplified. The relationship among common-rail pressure, fuel injection flow, fuel supply flow and fuel return flow can be written as:

$$\frac{\mathrm{d}P_{cr}}{\mathrm{d}t} = \frac{K}{V}(Q_1 - Q_2 - Q_3) \tag{1}$$

where Q_1 denotes the fuel supply flow of a HVCR pump, Q_2 denotes the injector's fuel injection flow, Q_3 denotes fuel return flow of the pressure control valve (PCV), P_{cr} denotes the fuel pressure in the HVCR pipe, K denotes the diesel's elastic modulus and V denotes the high-pressure common rail volume.

For accurately acquiring the quantitative relations among common rail pressure, the engine's rotational speed and fuel injection quantity, a 6L16/24 diesel was selected for the experiments on a HPCR testing platform. The fuel supply and return data at different common-rail pressures of 80MPa, 100MPa, 120MPa and 140MPa were recorded for further analysis.

Fuel supply flow mainly depends on the rotational speed of the high-pressure fuel pump which equals to half of the diesel's rotational speed n and the common rail pressure P_{cr}.Therefore, the following expression can be derived:

$$Q_1 = k_1 n + c_1 \tag{2}$$

Then, a quadratic polynomial of P_{cr} relative to k_1 and c_1 can be acquired through fitting, which can be written as:

$$k_1 = -0.0000000392 P_{cr}^{\ 2} + 0.00000273 P_{cr} + 0.001065 \tag{3}$$

$$c_1 = 0.0000581 P_{cr}^{\ 2} - 0.00356 P_{cr} - 0.072 \tag{4}$$

If the injector's fuel injection flow q is set as the circulating fuel injection flow, the following expression can be acquired:

$$Q_2 = q \tag{5}$$

The fuel return quantity increases with the rising of HPCR pressure. In this study, by neglecting the other factors of the equipment, fuel return quantity is directly determined by HPCR pressure:

$$Q_3 = 0.00416 P_{cr} \tag{6}$$

According to equations (1)-(6), the mathematical model of a HPCR nonlinear system can be described as:

$$\frac{\mathrm{d}P_{cr}}{\mathrm{d}t} = (a_1P_{cr}^2 + a_2P_{cr} + a_3)n + a_4P_{cr}^2 + a_5P_{cr} + a_6 + a_7q \tag{7}$$

where a_1=−0.00392, a_2=0.273, a_3=106.5, a_4=5.81, a_5=−35.6, a_6=−720, a_7=929.6

Obviously, the differential equation not only includes the nonlinear quadratic component of HPCR pressure but also connects with the diesel's rotational speed and fuel injection quantity.

The differential equation of the rotational speed control system is also required. Similarly, using curve fitting method, the circulating fuel injection quantity data and average indicating pressure p_i at different rotational speeds (900 r/min, 1000 r/min, 1100 r/min and 1200 r/min, respectively) were analyzed. The average indicating pressure can be expressed as:

$$p_i = k_2q + c_2 \tag{8}$$

Through quadratic polynomial fitting, the following expression can be derived:

$$k_2 = -0.0024025n^2 + 6.9n + 4873.3 \tag{9}$$

$$c_2 = 0.00003346n^2 - 0.107n - 71.5294 \tag{10}$$

The variation of diesel rotating speed directly depends on the difference between output torque M and load torque M_l. According to the D' Alembert fundamental equation of motion [6], the following expression can be acquired:

$$J\frac{\mathrm{d}w}{\mathrm{d}t} = M - M_l \tag{11}$$

Assuming that the characteristic of load torque M_l is the vehicle's stable road characteristic, the following expression can be derived:

$$M_l = l \cdot n^2 \tag{12}$$

Let l = 0.00007 and $J = 5kg \cdot m^2$, the nonlinear mathematic model for the diesel's rotational speed controlling can be written as:

$$\frac{\mathrm{d}n}{\mathrm{d}t} = b_1n^2 + b_2n + b_3 + (b_4n^2 + b_5n + b_6)q \tag{13}$$

where b_1=−0.0001, b_2=−0.144, b_3=−114.7, b_4=−0.003, b_5=8.4, b_6=5296

According to equations (7) and (13), the mathematic model of a nonlinear HPCR and speed control system can be written as:

$$\begin{cases} \dfrac{\mathrm{d}P_{cr}}{\mathrm{d}t} = (a_1P_{cr}^2 + a_2P_{cr} + a_3)n + a_4P_{cr}^2 + a_5P_{cr} + a_6 + a_7q \\ \dfrac{\mathrm{d}n}{\mathrm{d}t} = b_1n^2 + b_2n + b_3 + (b_4n^2 + b_5n + b_6)q \end{cases} \tag{14}$$

This model is characterized by the quadratic component with nonlinearity and includes two state variables to be controlled, namely, common rail pressure and the diesel's rotational speed.

2.2. Design of a sliding mode controller (SMC)

For the system as described in equation (14), the state variable can be set as:

$$x_1 = \int_0^t e_1\mathrm{d}t,\ \ x_2 = e_1 = P_{cr0} - P_{cr},$$

where P_{cr0} denotes the preset common rail pressure and e_1 denotes the pressure difference.

$$x_3 = \int_0^t e_2\mathrm{d}t,\ \ x_4 = e_2 = n_0 - n,$$

where n_0 denotes the preset value of the diesel's rotational speed and e_2 denotes the difference of rotational speed.

Therefore, the state equation of the system can be rewritten as:

$$\begin{cases} \dot{x}_1 = x_2 \\ \dot{x}_2 = -(a_1 P_{cr}^2 + a_2 P_{cr} + a_3)n - (a_4 P_{cr}^2 + a_5 P_{cr} + a_6) - a_7 q \\ \dot{x}_3 = x_4 \\ \dot{x}_4 = -(b_1 n^2 + b_2 n + b_3) - (b_4 n^2 + b_5 n + b_6)q \end{cases} \tag{15}$$

The state equation of a kind of affine nonlinear system can be written as:

$$\frac{\mathrm{d}}{\mathrm{d}t}\mathbf{x} = \mathbf{f}(\mathbf{x}) + \mathbf{g}(\mathbf{x})u \quad \mathbf{x} \in R^n \quad u \in R \tag{16}$$

where f(x) and g(x) are two function vectors

$$\begin{cases} \mathbf{f}(\mathbf{x}) = [f_1(\mathbf{x}) \ \ f_2(\mathbf{x}) \ \ \cdots \ \ f_n(\mathbf{x})]^T \\ \mathbf{g}(\mathbf{x}) = [g_1(\mathbf{x}) \ \ g_2(\mathbf{x}) \ \ \cdots \ \ g_n(\mathbf{x})]^T \end{cases}$$

where $f_i(\mathbf{x}) = f_i(x_1, \cdots, x_n)$, $g_i(\mathbf{x}) = g_i(x_1, \cdots, x_n)$, and $i = 1, \cdots, n$

This equation is linear for the control variable *u*. In a similar form, the system can be rewritten as:

$$\begin{cases} \dot{x}_1 = x_2 \\ \dot{x}_2 = -(a_1 P_{cr}^2 + a_2 P_{cr} + a_3)n - (a_4 P_{cr}^2 + a_5 P_{cr} + a_6) - a_7 u \\ \dot{x}_3 = x_4 \\ \dot{x}_4 = -(b_1 n^2 + b_2 n + b_3) - (b_4 n^2 + b_5 n + b_6)u \end{cases} \tag{17}$$

Therefore, we can derive that:

$$\mathbf{f}(\mathbf{x}) = \begin{bmatrix} f_1(\mathbf{x}) \\ f_2(\mathbf{x}) \\ f_3(\mathbf{x}) \\ f_4(\mathbf{x}) \end{bmatrix} = \begin{bmatrix} x_2 \\ -(a_1 P_{cr}^2 + a_2 P_{cr} + a_3)n \\ -(a_4 P_{cr}^2 + a_5 P_{cr} + a_6) \\ x_4 \\ -(b_1 n^2 + b_2 n + b_3) \end{bmatrix}$$

$$\mathbf{g}(\mathbf{x}) = \begin{bmatrix} g_1(\mathbf{x}) \\ g_2(\mathbf{x}) \\ g_3(\mathbf{x}) \\ g_4(\mathbf{x}) \end{bmatrix} = \begin{bmatrix} 0 \\ -a_7 \\ 0 \\ -(b_4 n^2 + b_5 n + b_6) \end{bmatrix}$$

where $u = q$, $P_{cr} = P_{cro} - x_2$, $n = n_0 - x_4$

Assuming $s = s(\mathbf{x}) = s(x_1, x_2, x_3, x_4)$ denotes the switching function in sliding-mode variable structure controlling, the following expression can be written as:

$$s(\mathbf{x}) = c_1 x_1 + c_2 x_2 + c_3 x_3 + x_4 \tag{18}$$

$$u_{eq}(\mathbf{x}) = -[\frac{\partial}{\partial \mathbf{x}} s(\mathbf{x})\mathbf{g}(\mathbf{x})]^{-1} \frac{\partial}{\partial \mathbf{x}} s(\mathbf{x})\mathbf{f}(\mathbf{x}) = \frac{c_1 x_2 - c_2 A_1 n - c_2 A_2 + c_3 x_4 - c_4 B_1}{-c_2 a_7 - B_2} \tag{19}$$

where $A_1 = a_1 P_{cr}^2 + a_2 P_{cr} + a_3$, $A_2 = a_4 P_{cr}^2 + a_5 P_{cr} + a_6$, $B_1 = b_1 n^2 + b_2 n + b_3$, $B_2 = b_4 n^2 + b_5 n + b_6$

By substituting equation (19) into equation (16), the sliding-mode motion equation can be described as:

$$\begin{cases} \frac{\mathrm{d}}{\mathrm{d}t}\mathbf{x} = \mathbf{f}(\mathbf{x}) - \mathbf{g}(\mathbf{x})[\frac{\partial}{\partial \mathbf{x}} s(\mathbf{x})\mathbf{g}(\mathbf{x})]^{-1} \frac{\partial}{\partial \mathbf{x}} s(\mathbf{x})\mathbf{f}(\mathbf{x}) \\ s(\mathbf{x}) = 0 \end{cases} \tag{20}$$

SMVSC can be solved by reaching law control. In this study, the simplest constant reaching law was used in the solution which is $ds/dt = -\varepsilon \operatorname{sgn}(s), \quad \varepsilon > 0$. Therefore, the reaching law can be rewritten as:

$$-\varepsilon \operatorname{sgn}(s) = \sum_{i=1}^{n} c_i f_i(\mathbf{x}) + [\sum_{i=1}^{n} c_i g_i(\mathbf{x})]u \tag{21}$$

The control rate u can be solved according to Eq. (21):

$$\begin{cases} u^+ = \dfrac{\varepsilon + c_1 x_2 - c_2 A_1 n - c_2 A_2 + c_3 x_4 - c_4 B_1}{-c_2 a_7 - B_2} & s > 0 \\ u^- = \dfrac{-\varepsilon + c_1 x_2 - c_2 A_1 n - c_2 A_2 + c_3 x_4 - c_4 B_1}{-c_2 a_7 - B_2} & s < 0 \end{cases} \tag{22}$$

3. Software simulations

The diesel engine's rotational speed was controlled by a SMVSC. Firstly, based on the identified deviation of rotational speed, the control input u was obtained, i.e., the circulating fuel injection quantity q (g). Secondly, using both the circulating fuel injection quantity and the feedback rotational speed as the known quantities, the average indicating pressure can be acquired by looking up the MAP charts [7], and the output torque was calculated by subtracting the average friction pressure from the average indicating pressure. Finally, according to D' Alembert principle, the actual rotational speed was acquired based on the difference between the output torque M and the load torque M_l.

As to the controlling of HPCR pressure, the control input u was firstly acquired based on the pressure deviation. Then, both the control input u and the feedback pressure were adopted as the known quantities, and fuel injection pulse width, the opening degree of PCV valve and the actual HPCR pressure were obtained by looking up the MAP charts of the common rail pressure of the diesel engine. The simulation program of the system is shown in Figure 1. Simulation results of the common rail pressure and speed of the diesel engine are shown in Figures 2-7.

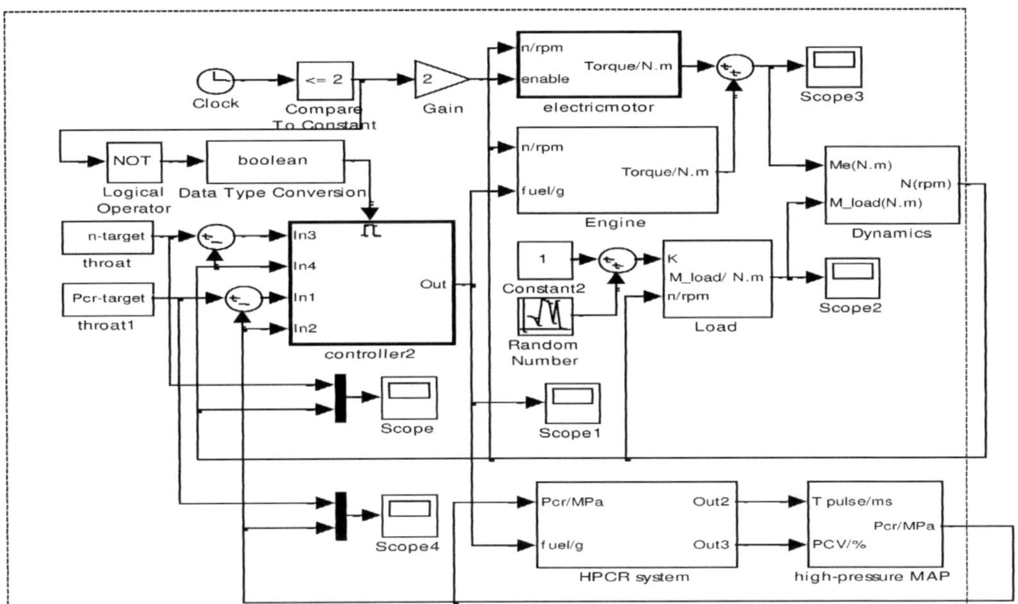

Figure 1. Simulation structure figure of the common rail and speed system.

IOP Conf. Series: Journal of Physics: Conf. Series **1016** (2018) 012017 doi:10.1088/1742-6596/1016/1/012017

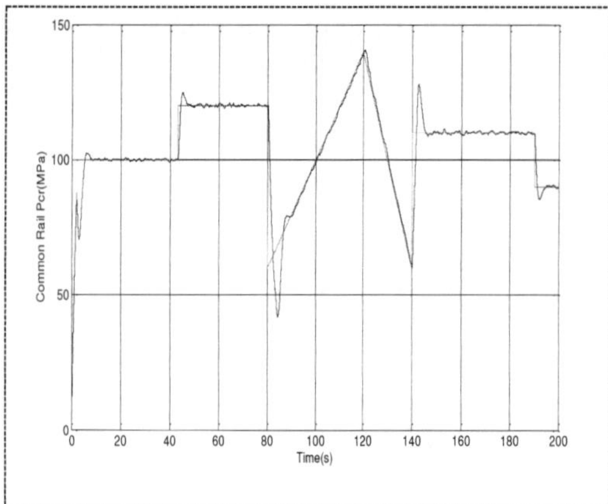

Figure 2. Common rail pressure simulation of PID.

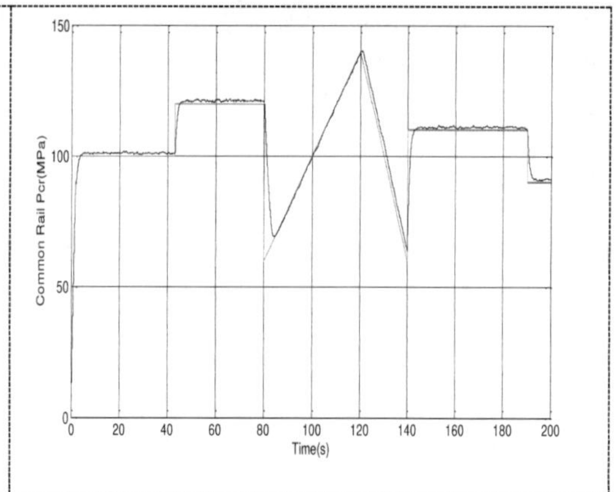

Figure 3. Common rail pressure simulation of SMVSC.

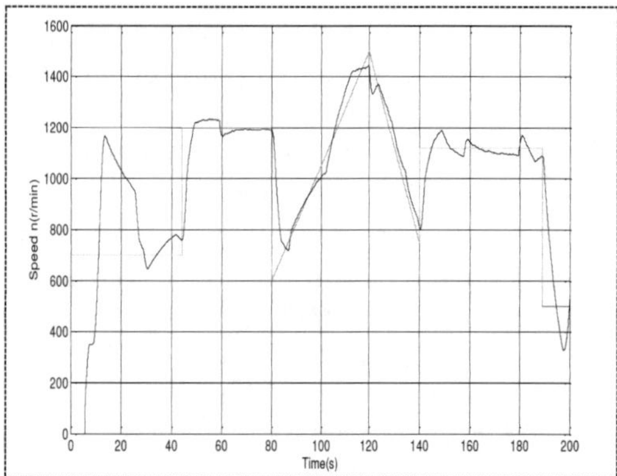

Figure 4. Rotational speed simulation of PID.

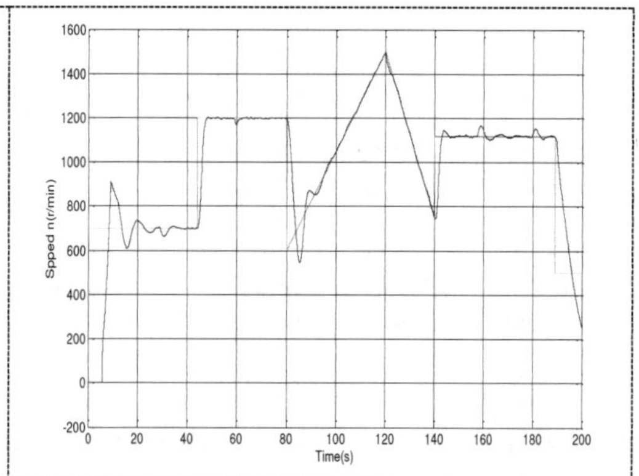

Figure 5. Rotational speed simulation of SMVSC.

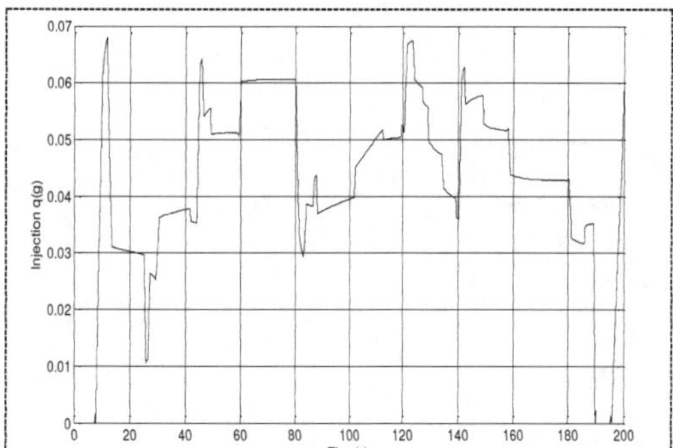

Figure 6. Variation of oil injection with PID.

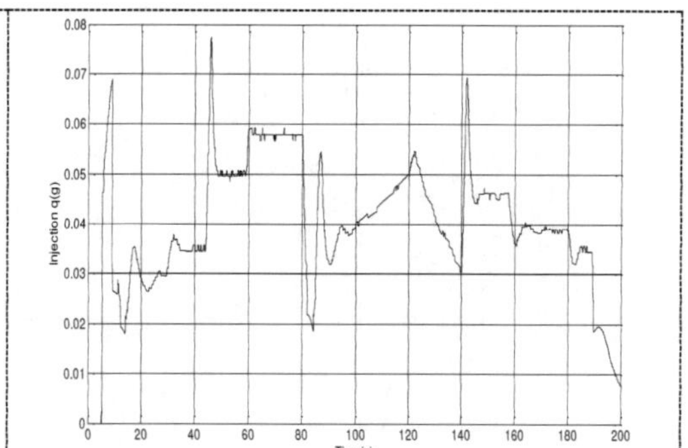

Figure 7. Variation of oil injection with SMVSC.

According to the simulation results of the vehicle's stable road characteristic, For SMVSC, the rotational speed overshoot was much smaller than PID control, especially in the initial stage of diesel engine starting, and speed fluctuated within 10 r/min in the stable stage. Additionally, the diesel engine's transient speed-regulating rate was smaller than 10%, the stable speed-regulation rate was about 0%, and the fluctuation ratio of rotational speed was smaller than 1%. Under a sudden load-up of 100%, the stabilization time of 5% error was smaller than 4 seconds.

For the common rail pressure control, SMVSC varies within a range of 5%. The simulation results fully proved the feasibility of variable structure control theory in investigating the common rail and rotational speed systems of nonlinear diesel engine. Using variable structure control method, the system's nonlinear characteristics were well maintained.

4. Conclusions

In previous studies, the diesel engine's rotational speed, common rail pressure, fuel viscosity temperature and cooling water temperature were generally controlled by traditional PID strategy. Although the control structure was simple, the control precision, stability, robustness and control performances were far from ideal.

The fundamental difference between sliding-mode variable structural control and conventional control lies in the control's discontinuity, i.e., a kind of switching characteristics that can make the system structure change at any time. The control characteristics can force the system structure to change purposefully in a leaping mode during the transient process according to the system state at that moment (deviation and its all-order derivatives), and make up-and-down motions slightly but frequently along the specified state trajectory under certain conditions. Using sliding model variable structure control method, system shows a series of advantages including fast response, small overshoot and strong robustness.

References

[1] Bahram Bahri and Azhar Abdul Aziz 2013 Misfiring cycle pressure measurement for diesel-converted HCCI engine *2013 3rd International Conference on Instrumentation Control and Automation* pp 45-50

[2] An Xiaohui and Liu Bolan 2010 Electronic control unit development for unit pump diesel engine *2010 International Conference on Optoelectronics and Image Processing* vol 1 pp 479-482

[3] Farouk N and Leghmizi 2012 Speed control system on marine diesel engine based on a self-tuning fuzzy PID controller *Research Journal of Applied Sciences, Engineering and Technology* vol 4 chapter 6 pp 686-690

[4] Xu F and Chen H 2015 Fast Nonlinear model predictive control on FPGA using particle swarm optimization *IEEE Transactions on Industrial Electronics* vol 63 chapter 5 pp 310-321

[5] Fakhari V and Ohadi A 2013 A robust adaptive control scheme for an active mount using a dynamic engine model *Journal of Vibration & Control* vol 21 chapter 11 pp 1-23

[6] Sun Z and Kuo T W 2010 Transient control of electro-hydraulic fully flexible engine valve actuation system *IEEE Transactions on Control Systems Technology* vol 18 chapter 3 pp 613-621

[7] Wang Y and Chen B 2011 Simulation research on intelligent control for propulsion system of large ship *Oceans IEEE* vol 5 chapter 7 pp 1-7

Acknowledgements

This research project has been supported by the independent innovation research fund project (2017-ND-A1-03) of Wuhan University of Technology. I'd like to express my gratitude.

IOP Conf. Series: Journal of Physics: Conf. Series **1016** (2018) 012018 doi:10.1088/1742-6596/1016/1/012018

Workspace Design for 6 Degree of Freedom Machining Bed

M F Shah[1], Z Kausar[2,*] and S S Farooq[1]

1. Department of Mechanical Engineering, Khwaja Fareed University of Engineering and Information Technology, Abu Dhabi Road, Rahim Yar Khan, Pakistan
2. Department of Mechatronics Engineering, Air University, E-9, Islamabad, Pakistan

[*]Email: zkau001@aucklanduni.ac.nz

Abstract. Machining is a process in which raw material is cut into desired shape and size using different material removal processes. The accuracy and precision of workpiece is effected by present day high speed spindle technologies. To deal with such a problem it is proposed to replace current machining techniques with a 6 DOF machining bed which gives motion to workpiece in six degrees whereas the tool remains static. An issue in using such machining beds is workspace due to limitation of its joints movement and this should be known. This paper presents an algorithm for the determination of workspace for such machining beds. The algorithm is implemented in simulation which verifies the suitability of bed for machining in 6 degree of freedom.

1. Introduction

The process of removal of material from a workpiece in order to obtain the desired shape and dimensions can be termed as machining. Considered to be a manufacturing part for metals machining can also be done on ceramics, composites and wood. Shaping, sawing, broaching, turning and drilling all can be categorized as types of machining [1]. The process of machining comes into play where high accuracy and precision are required. Conventional machining technologies use high speed tools which allow the material to remove very quickly. But these high speed tools are vulnerable to internal and external vibrations which may result in poor precision and low accuracy[2].Current CNC technologies have multi tool axis, but the accuracy of these machines is influenced by various types of errors and tool wear[3,4]. Such problem can be solved by using a mechanism in which the tool remains static and the workpiece moves accordingly. A parallel machining bed able to give rotation to workpiece in 6 DOF was proposed by Shah et al [5] and Kausar et al [2]in their research works. These types of parallel manipulators are widely known as Stewart Platform. In such manipulators shown in figure 1 the top plate gives motion in 6 degrees[6,7].These parallel manipulators give high stiffness and have high load bearing capacities [8] hence have gained attention from researchers and industries as well.

A challenge in the proposed system is the limitation of size of the workpiece. It is very important before starting this process of machining that the volume within which the machining bed can reach for given shape of the workpiece. Workspace of a parallel manipulator is of much concern while designing it. From both theoretical and industrial point of view the optimum design of such parallel manipulators can make them of more use [9]. The volume within which the end effector of the parallel manipulators can reach is the basic requirement for their practical usage [10]. Collection of all points $\{x\ y\ z\}^T$ where the robot can reach in at least one of its orientation is known as its reachable workspace [3]. The workspace of parallel manipulators is limited [11]because of limitations on its joints, link interface and limited range of link lengths. Keeping in view above mentioned issues workspace for each length of parallel manipulators is calculated separately using inverse kinematics [10] but this research proposes an algorithm to calculate the cumulative workspace for the machining bed. The paper analyzes the workspace of a 6 DOF parallel manipulator used for the machining purpose by first solving the inverse kinematics presented in section 2 and later simulating the workspace using proposed algorithm with and without constraints.

2. Kinematic Modeling

The problem is stated as a position of the end effector of the manipulator is known to us depending upon the coordinates of the workpiece, respective leg lengths are required to be determined. For this purpose inverse kinematics of the parallel machining bed was used.

From figure 1 it can be seen that two plates of machining bed are connected through six legs with variable lengths. The base plate of the machining bed was considered as the reference frame[12]. There are many orientations of legs in which these platforms are proposed[13–17], we here propose that the legs along with the ball and socket joint are attached at an equal angle of 60 degrees with the spherical top plate whereas at the bottom these lengths along with the universal joints are place in form of pairs at equidistance from each other and other pair of legs as well.

A schematic view of the ith leg of parallel machining bed is shown in figure 2. L_i is the length of ith leg of machining bed, T is the translational vector, P_i is the point which defines the upper anchor point from the center of the movable top plate and B_i is the point which defines the anchor point from the center of the reference bottom plate. $G_i$is the resultant vector obtained as a result of head to tail rule addition [3]. Equation obtained for i^{th} leg of the machining bed by using geometrical relations and vector addition is (1):

$$\overrightarrow{L_i} = \vec{T} + {}^{P}_{B}R\overrightarrow{P_i} - \overrightarrow{B_i} \tag{1}$$

Where ${}^{P}_{B}R$ is rotation matrix from O_B to O_P. Rotation matrix was designed using (2):

$${}^{P}_{B}R = R_Z(\psi) + R_Y(\theta) + R_X(\varphi) \tag{2}$$

$$= \begin{pmatrix} c(\psi)\,c(\theta) & -s(\psi)\,c(\varphi) + c(\psi)\,s(\theta)\,s(\varphi) & s(\psi)\,s(\varphi) + c(\psi)\,s(\theta)\,c(\varphi) \\ s(\psi)\,c(\theta) & c(\psi)\,c(\varphi) + s(\psi)\,s(\varphi)\,s(\theta) & -c(\psi)\,s(\varphi) + s(\psi)\,s(\theta)\,c(\varphi) \\ -s(\theta) & c(\theta)\,s(\varphi) & c(\theta)\,c(\varphi) \end{pmatrix} \tag{3}$$

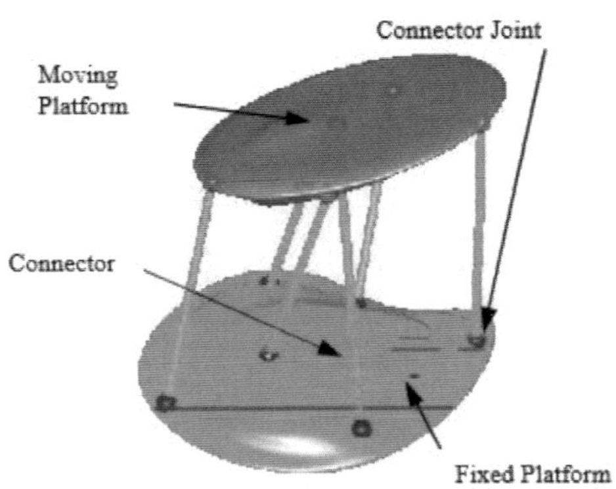

Figure 1. Configuration of a Stewart Platform [5]

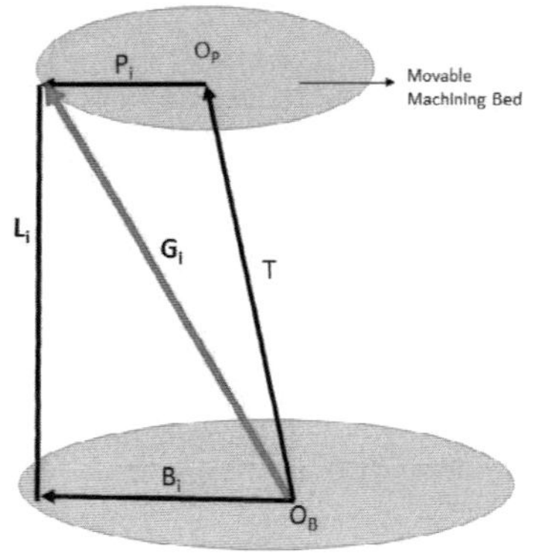

Figure 2. Schematic View of ith Leg of Parallel Machining Bed

In (3) and in rest of this paper c is used as abbreviation for cosine and s for sine, whereasψ, θ and φ are angels in respective axis.Distribution of angels between six legs of the machining bed at the top plate as well as the bottom plate are shown in figure 3. Initially it was assumed that legs translate ***a*** units in z axis only. Keeping in view this assumption the T vector was written as shown in (4):

$$\vec{T} = \begin{pmatrix} 0 \\ 0 \\ \boldsymbol{a} \end{pmatrix} \tag{4}$$

Using the data obtained in (4) and interpreting the values of Pi and Bi from figure 3 and figure 4, (1)was solved for each of the six legs of the machining bed. Resulting expressions are listed in (5) to (10).

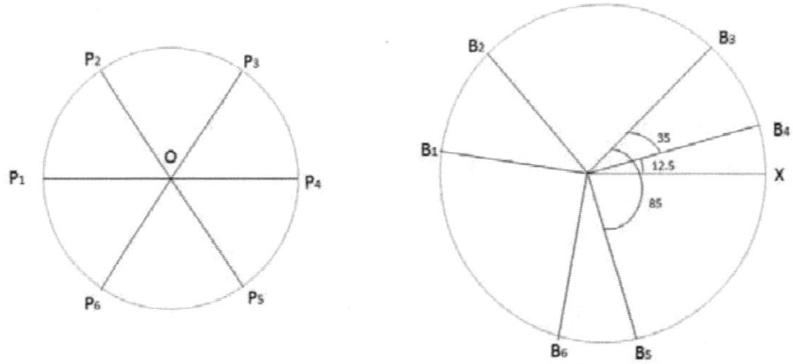

Figure 3. Angle Distribution on Movable Top Plate (Left) and Bottom Plate (Right) of Machining Bed.

$$L_1 = \begin{pmatrix} rc\psi c\theta - Rc12.5 \\ rs\psi c\theta - Rc12.5 \\ a - rs\theta \end{pmatrix} \tag{5}$$

$$L_2 = \begin{pmatrix} rc60c\psi c\theta - rs60s\psi s\varphi + rs60c\psi s\theta s\varphi - Rc47.5 \\ rc60s\psi c\theta + rs60c\psi c\varphi + rs60s\varphi s\psi s\theta - Rs47.5 \\ a - rc60s\theta + rs60c\theta s\varphi \end{pmatrix} \tag{6}$$

$$L_3 = \begin{pmatrix} -rc60c\psi c\theta - rs60s\psi s\varphi + rs60c\psi s\theta s\varphi + Rc47.5 \\ -rc60s\psi c\theta + rs60c\psi c\varphi + rs60s\varphi s\psi s\theta - Rs47.5 \\ a + rc60s\theta + rs60c\theta s\varphi \end{pmatrix} \tag{7}$$

$$L_4 = \begin{pmatrix} -rc\psi c\theta + Rc12.5 \\ -rs\psi c\theta - Rs12.5 \\ a + rs\theta \end{pmatrix} \tag{8}$$

$$L_5 = \begin{pmatrix} -rc60c\psi c\theta + rs60s\psi c\varphi + rs60c\psi s\theta s\varphi + Rc17.5 \\ -rc60s\psi c\theta - rs60c\psi c60 - rs60s\varphi s\psi s\theta + Rs17.5 \\ a + rc60s\theta - rs60c\theta s\varphi \end{pmatrix} \tag{9}$$

$$L_6 = \begin{pmatrix} rc60c\psi c\theta + rs60s\psi c\varphi - rs60c\psi s\theta s\varphi - Rc17.5 \\ rc60s\psi c\theta - rs60c\psi c60 - rs60s\varphi s\psi s\theta + Rs17.5 \\ a - rc60s\theta - rs60c\theta s\varphi \end{pmatrix} \tag{10}$$

Leg lengths are determinable knowing the orientation of the center of the top plate of the machining bed where the workpiece is to be moved, T depending upon the shape of the workpiece, rotation matrix (3) and P_i and B_i .

3. Workspace

IOP Conf. Series: Journal of Physics: Conf. Series **1016** (2018) 012018 doi:10.1088/1742-6596/1016/1/012018

3.1. Algorithm
A number of techniques are proposed to determine workspace margins for a manipulator. These techniques include analytical methods, iterative methods and numerical methods. To design workspace for the machining bed a MATLAB ® code is developed by using leg lengths obtained using (5)-(10). The algorithm of the workspace design is as given in figure 4.

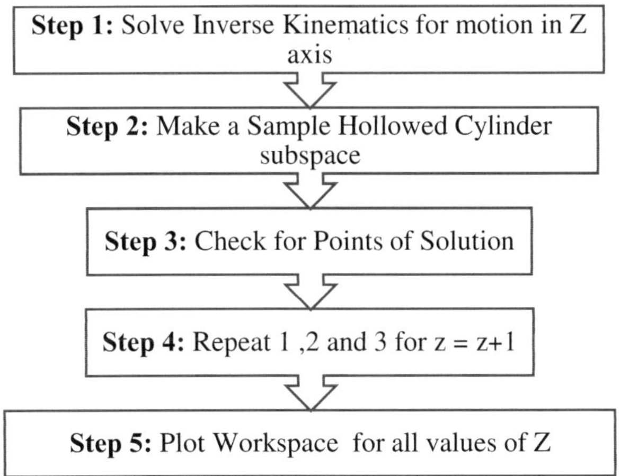

Figure 4. Workspace design algorithm for the 6 DoF machining bed.

3.2. Simulation results
The algorithm proposed in section 3.1 is simulated for the machining bed in Matlab. For each leg and each angle a step size of 0.4 is given from the interval -1 to 1 radians. For each sliding movement, **a**, the program is run for the step size of 20 from the interval of 0 to 800mm. Then the mapping of leg lengths of the machining bed obtained through inverse kinematics is obtained against the angles to demonstrate the movement of the machining bed. Finally graphics was added for separate mapping and enhanced understanding of each leg as shown in figure 4. In second test all the parameters were kept as in first case but a geometric constraint was added to restrict the machining bed kinematics (Leg Lengths) between 400 to 750mm.The resulting dexterous workspace obtained for the machining bed is shown in figure 5.The calculated work space is for the whole platform, including legs and movable top plate. Any change in the parameters will result in change of the work space. The work space obtained is of zig-zag shape. So, machining bed will work for any work piece falling within limits of constraints applied.

4. Conclusions
An algorithm is proposed for design of workspace for a six degree of freedom machining bed. The workspace requirement may change with the change in shape and size of the workpiece. The proposed algorithm draws a cylinder (boundary conditions) knowing maximum size of leg actuators and then draws trajectory of the workpiece and confirms whether the workpiece shape lies within boundary condition or not. The same algorithm may be applied to design parameters of legs knowing the maximum requirement of the workpiece machining. Future work may include the control of workspace and position of the workpiece.

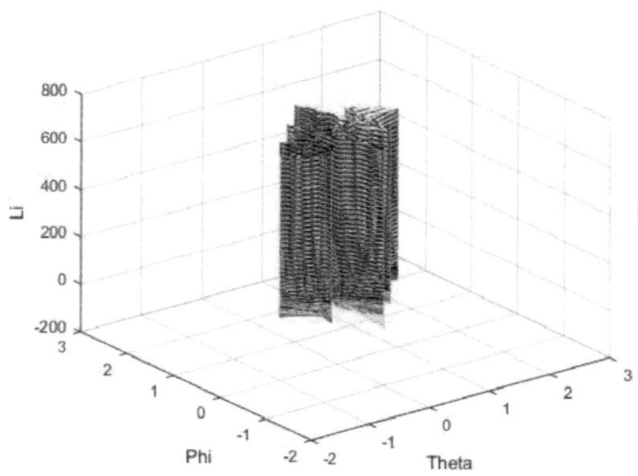

Figure 5. Workspace of a 6 D.O.F Machining Bed

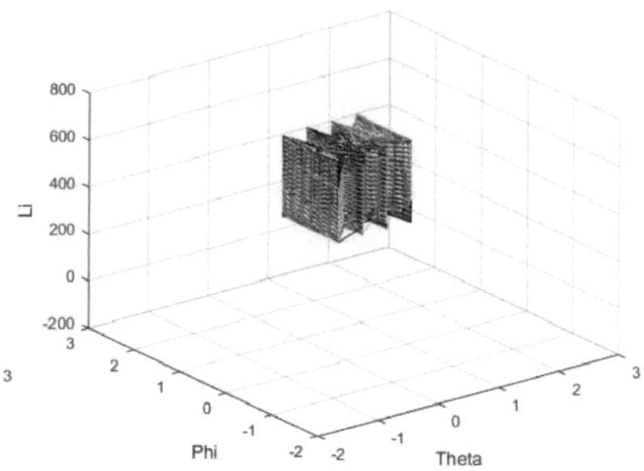

Figure 6. Dexterous Workspace for 6 D.O.F Machining Bed

References

[1] Sutherland J . 2017 *Lecture Notes on Manufacturing* (Houghton)
[2] Irshad M A 2014 A Parallel Robotic Mechanism Replacing a Machine Bed for Micro-Machining 228–33
[3] Cheng Q, Sun B, Zhao Y and Gu P 2016 A method to analyze the machining accuracy reliability sensitivity of machine tools based on Fast Markov Chain simulation *Eksploat. i Niezawodn. - Maint. Reliab.* **18** 552–64
[4] Zhang Z, Liu Z, Cheng Q, Qi Y and Cai L 2017 An approach of comprehensive error modeling and accuracy allocation for the improvement of reliability and optimization of cost of a multi-axis NC machine tool *Int. J. Adv. Manuf. Technol.* **89** 561–79
[5] Shah M F, Kausar Z and Durrani F K 2016 Design , Modeling and Simulation of Six Degree of Freedom Machining Bed **53** 163–76
[6] Wang Y 2008 Symbolic Kinematics and Dynamics Analysis and Control of a General Stewart Parallel Manipulator
[7] Ridgeway S C and Iii C D C Control Considerations in the Design of a Parallel Kinematic Machine with Separate Actuation and Metrology Mechanisms Center for Intelligent Machines and Robotics Department of Mechanical and Aerospace Engineering P . O . Box 116300 University of Florida
[8] Lee S, Song J, Choi W and Hong D 2003 Position control of a Stewart platform using inverse dynamics control with approximate dynamics **13** 605–19
[9] Staicu S and Zhang D 2008 Dynamic modelling of a 4-DOF parallel kinematic machine with revolute actuators *Int. J. Manuf. Res.* **3** 172–87
[10] Shao Z, Tang X, Wang L and Sun D 2014 Atlas based kinematic optimum design of the Stewart parallel manipulator *Chinese J. Mech. Eng.* **28** 20–8
[11] Shah M F, Kausar Z, Durrani F K, Tahir A H and Khalid N 2016 Design and Development of Six Degrees of Freedom Parallel Mechanism for the Purpose of Machining
[12] Bingul Z and Karahan O Dynamic modeling and simulation of Stewart platform, Serial and parallel robot manipulators–kinematics, dynamics, control and optimiziation, Ed. by S. *ISBN 978-953-51-0437-7*
[13] Pedrammehr S, Reza M, Qazani C and Rahmani A Forced vibration analysis of milling machine ' s hexapod table under machining forces 1–15
[14] Tahboub K 2017 A Novel Three Degree of Freedom Stewart-like Platform for Testing Humanoid Postural Capabilities A Novel Three Degree of Freedom Stewart-like Platform for Testing
[15] Kim J, Park F C, Ryu S J, Kim J, Hwang J C, Park C and Iurascu C C 2001 Design and Analysis of a Redundantly Actuated Parallel Mechanism for Rapid Machining **17** 423–34
[16] Garagić D and Srinivasan K 2004 Contouring control of Stewart platform based machine tools *American Control Conference, 2004. Proceedings of the 2004* vol 4pp 3831–8
[17] Zeng J G, Liu J Y and Yu Q 2014 *Design and Development of 6DOF Hydraulic Motion Platform for Vehicle Driving Simulator* vol 505–506

IOP Conf. Series: Journal of Physics: Conf. Series **1016** (2018) 012019 doi:10.1088/1742-6596/1016/1/012019

Computational Model for Impact-Resisting Critical Thickness of High-Speed Machine Outer Protective Plate

Huaying Wu1, Li Zhong Wang1, Yantao Wang2, Xiaolei Yuan2

1School of Mechanical Engineering, Xi'an Jiaotong University,Xi'an,710049;
2Kocel Machine Tool Accessories Limited,Yinchuan, 750021

Email: wuhy@xjtu.edu.cn

Abstract. The blade or surface grinding blade of the hypervelocity grinding wheel may be damaged due to too high rotation rate of the spindle of the machine and then fly out. Its speed as a projectile may severely endanger the field persons. Critical thickness model of the protective plate of the high-speed machine is studied in this paper. For easy analysis, the shapes of the possible impact objects flying from the high-speed machine are simplified as sharp-nose model, ball-nose model and flat-nose model. Whose front ending shape to represent point, line and surface contacting. Impact analysis based on J-C model is performed for the low-carbon steel plate with different thicknesses in this paper. One critical thickness computational model for the protective plate of high-speed machine is established according to the damage characteristics of the thin plate to get relation among plate thickness and mass, shape and size and impact speed of impact object. The air cannon is used for impact test. The model accuracy is validated. This model can guide identification of the thickness of single-layer outer protective plate of a high-speed machine.

1. Introduction

Some hazardous factors such as splashing cuttings and high-speed flying tool or grinding wheel caused by accidental damage in high-speed cutting may exist (see Fig. 1), which will severely injure persons and even lead to death accidents, so the protective plate of the high-speed machine should resist certain impact damage and secure operators around the high-speed machine.

Now the impact characteristics of a thin plate are mainly studied in military, aerospace, earth and planetary physics field. The impact characteristics at a hypervelocity are focused (several km to tens of km per second) [1]. It is generally believed that the linear velocity of 150m per second is the dividing line between high speed and super high speed in machine tool. After the thin plate is impacted at a high speed, some debris cloud will be formed[2]. Zhang Yongqiang, Gai Fangfang, Liu Yingfang, at al respectively studied characteristics of debris cloud characters caused by the ball projectile, cylinder projectile and conical projectile[3-5]. Chi et al studied the debris cloud characteristics after the aluminum ball impacts 0.5mm and 1mm aluminum plate at a hypervelocity[6]. When the speed is lower, the thin plate will be torn and damaged. Rusinek et al studied penetration damage of low-carbon steel thin plate [7]. Gang Jingbo analyzed damage of the thin plate under action of the contact explosion load. The breach of the plated grillage shows the petal damage [8]. Du Zhipeng et al studied the model of the thin plate pedal damage under impact of a sharp-nose projectile, which can deduce the ballistic limit[9]. The damage styles of the thin plate are very different under

IOP Conf. Series: Journal of Physics: Conf. Series **1016** (2018) 012019 doi:10.1088/1742-6596/1016/1/012019

high-speed and hypervelocity impact. The outer protective thin plate of the high-speed machine is mainly damaged due to tearing instead of debris cloud.

a) Thrown blade in face mill b) A broken high-speed grinding wheel

Figure.1. Some tool accidents in high-speed machine

Based on research conclusion on the thin plate's high-speed impact, researchers have proposed Whipple protective structure and multiple enhanced protective structures based on it and analyzed ballistic limit characteristics, including density-grade thin plate1, sandwich cellular interlayer structure[10], corrugated sandwich protective structure[11], stainless steel mesh/aluminum plate protective structure[12], etc. The protective thin plate for hypervelocity impact is mainly made of the aluminum plate. Now the researched materials cover C/Si material[13], polyester plate[14], silk/epoxy/foam sandwich material[15], compound elastic materials[16], foam aluminum[17, 18], porcelain aluminum plate[19], etc. Generally, these structures and materials are used for hypervelocity impact cases. Although they can be referred in machine protection, the protection speed is too high and the cost is too high, so they are not fully suitable for civil high-speed machine field.

This paper takes the boundary velocity 150m/s of super high-speed as the studying speed for high-speed machine protection. This speed is approximately equivalent to the projectile speed, so it is very dangerous for operators close to the machine. Existing single-layer ballistic limit computing equation is based on the hypervelocity impact research[20], including five relational model between single-layer plate and impact object such as Fish-Summers, Schmidt-Hoisapple, Rockwell, JSC and correction equation[21], the computing equation is complicated and the key materials such as aluminum alloy and 304 stainless steel[22] are included. Now the high-speed machine protective plate is mainly made of Q235/SPCC. Thickness mainly depends on experiences. The supporting theory for thickness of impact-resisting critical protective plate is not provided. By analyzing impact resistance of the protective thin plate, one new simple and easy-to-calculate impact-resisting critical thickness model of the protective plate is established and theoretical support for design of the outer protective plate of the high-speed machine is available in this paper.

2. High-speed impact object model

Any impact object of the high-speed machine will contact the outer protective plate in a point, line and plane manner and the front impact will damage the plate much. The heavier 42g single grinding wheel in the measured CBN grinding wheel is used as the mass of the impact object. The end of the impact object is simplified as ball-nose, flat-nose and sharp-nose. Three impact objects are designed (refer to the figure 2). They are named as ball-nose projectile, flat-nose projectile and sharp-nose projectile according to the end characteristics to analyze protection characteristics of the protective plate under different impact objects.

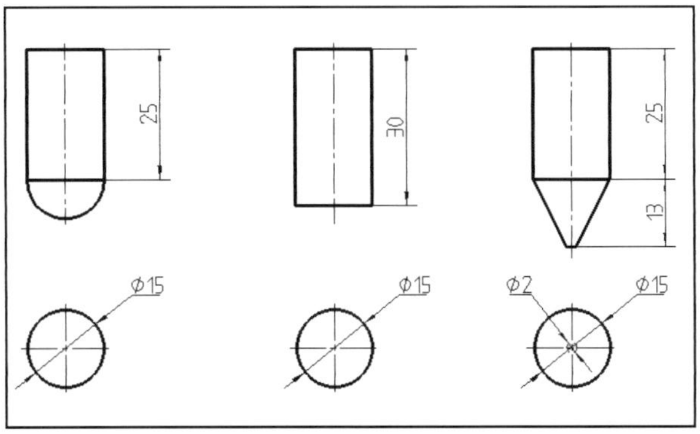

Figure.2 Three different types of impact objects

3. Simulation model of high-speed impact

ABAQUS display mechanics analysis module is used for simulation in this paper. The impact object is set as the rigid body. The Johnson-Cook material failure model is used[23] to analyze the heat insulation instantaneous dynamics analysis in case of high-speed impact. The ultra-high line speed 150m/s[24] in the machine field is used as the impact speed. The common Q235 steel plate is used as the protective plate for protection analysis of the machine's outer plate.

3.1. Impact performance of three impact objects with different shapes on one protective plate

Two-millimeter-thick Q235 steel plate is used as the protective steel plate. Shown as the figure 3, when the ball-nose projectile impacts the plate, the plate will be gradually deformed. The maximal impact force occurs at the maximal diameter to break the plate. The diameter of the broken chip is similar to the maximal diameter of the projectile body. The damaged hole edge of the plate shows slightly visible petal. When the flat-nose projectile impacts the plate, the surrounding shape edge will form the blanking effect and the round chip with the diameter as the projectile diameter will be directly extruded and chiseled, so the extrusion and chiseling damage occurs. When the sharp-nose projectile pierces through the protective plate, no materials fall off and it leads to petal piercing damage. When the protective plate is impacted by the sharp-nose projectile, the plate will be quickly damaged due to local stress concentration. With advance of the projectile, the initial stress wave will generated the circumferential and radial high tension stress in the plate. When the circumferential stress reaches the extreme stress, the cracking conditions are satisfied and some tenacious cracks will be generated. The cracks will continuously expand along the initial direction to form the bigger crack and finally form the petal damage.

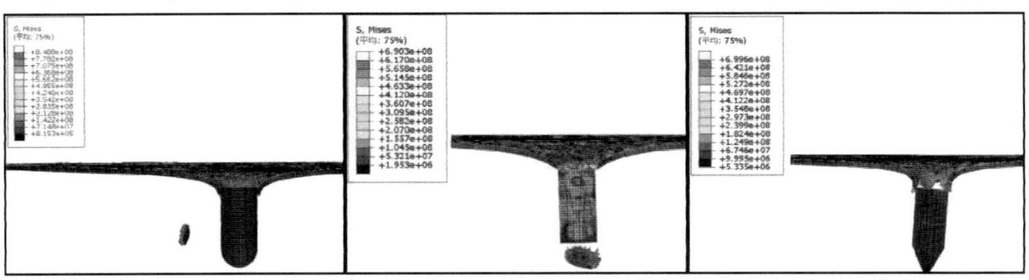

(a) Ball-nose projectile (b) Flat-nose projectile (c) Sharp-nose projectile

Figure.3. Phenomena of three projectiles impact on protective plate

By analyzing speed change curve of three objects before and after piercing through the protective plate (see the figure 4), when the ball-nose projectile, flat-nose projectile and sharp-nose projectile pierce through the protective plate, the speed loss is 33.7%, 62.4% and 31.8%. The time for full penetration is 0.2ms, 0.25ms and 0.3ms respectively. The speed loss of the flat-nose projectile reaches the maximum. The speed loss of the ball-nose projectile is similar to it of the sharp-nose projectile.

The speed reduction curve of the sharp-nose projectile and ball-nose projectile is similar to the sectional bus line of the head. The speed of the sharp-nose projectile will linearly reduce. The speed of the ball-nose projectile will reduce as a curve. The speed of the ball-nose projectile and sharp-nose projectile will change stably. The flat-nose speed experiences skip change, which may be caused by instantaneous cutting of the round chip. It will lead to different scraping between the section burr of the plate and flat-nose projectile. When the flat-nose projectile pierces through the plate, unstable friction resistance will lead to speed fluctuation. Based on above analysis, if the contact circumference between the impact object and protective plate is longer, it will significantly resist.

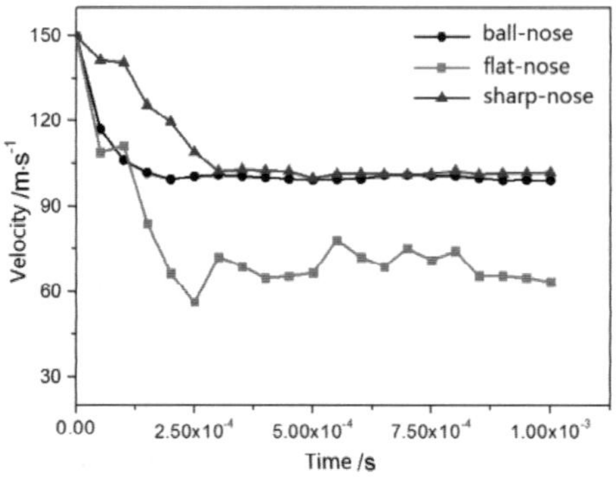

Figure.4. Speed change rule of different projectile impacts on 2mm protective plate

3.2. Impact performance of three impact objects with different thicknesses

Q235 plate is used and the proportional method is used to increase the plate thickness and analyze impact under different thicknesses.

3.2.1. Impact performance of ball-nose projectile on protective plates with different thicknesses. he increased thickness will significantly enhance damage resistance of the protective plate. Shown as the figure 5, when t is 2mm, the projectile will pierce through the plate. When t is 3mm, although the plate breaks, the impact-incurred broken chips do not fully separate from the protective plate, and the projectile does not pierce through the plate and rebound. When t is 4mm, the protective plate only has plastic deformation to form the tray concave, but it does not break and prevent against impact from the impact object. The thin plate damage under the high-speed projectile is similar to the stamping and flanging.

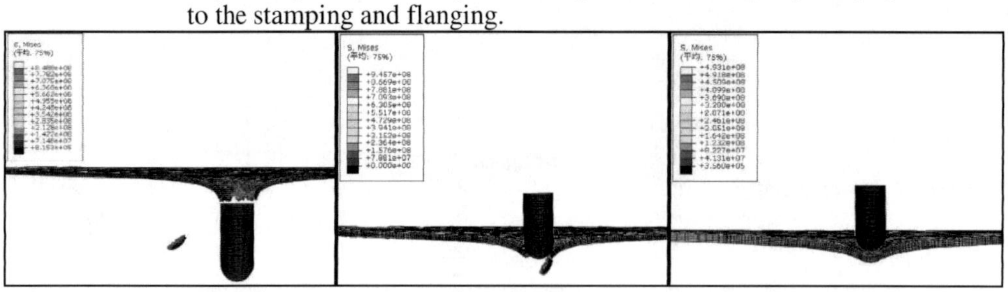

(a) t=2mm (b) t=3mm (c) t=4mm

Fig.5 Phenomena of ball-nose projectile impacts on different thickness protective plate

3.2.2. Impact performance of flat-nose projectile on protective plates with different thicknesses

Shown as the figure 6, when t is 2mm, the projectile will cut the plate and pierce through it. When t is 3mm, the plate surface suffers from slight damage and the plate will slightly rebound. The reverse speed is 2.58m/s. When t is 4mm, the plate is not fully damaged and bigger rebounding occurs to prevent against impact of flat-nose projectile. When the thickness of the protective plate reaches 3mm, the plate can prevent against impact of the flat-nose projectile. The thin plate damage under the flat-nose projectile is very similar to cracking of the plate blanking.

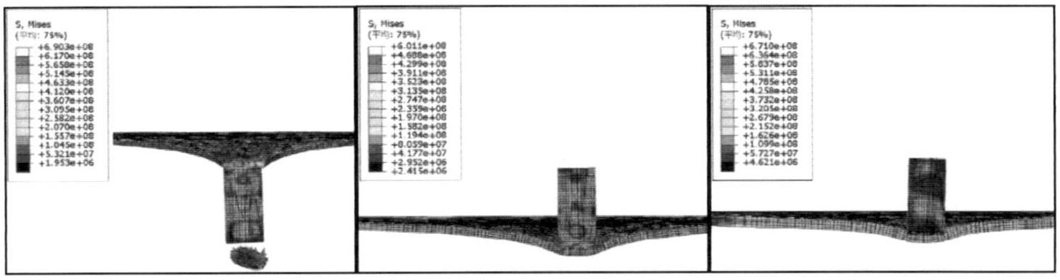

(a)t=2mm (b)t=3mm (c)t=4mm

Figure.6. Phenomena of flat-nose projectile impacts on different thickness protective plate

3.2.3. Impact performance of sharp-nose projectile on protective plate with different thicknesses

For the impact characteristics of the sharp-nose projectile protective plates with different thicknesses, refer to the figure 7. When t is 2mm, the projectile will pierce through the protective plate. When t is 3 and 4mm, the front part of the plate will be broken by the tip of the projectile, but the projectile does not pierce through the plate. The piercing depth of the tip becomes smaller and smaller. For the protective plate, the impact object does not fly out from the protective plate, namely the protective plate can protect. Under the high-speed impact of the sharp-nose projectile, the thin plate damage is similar to it of the stamping and flanging.

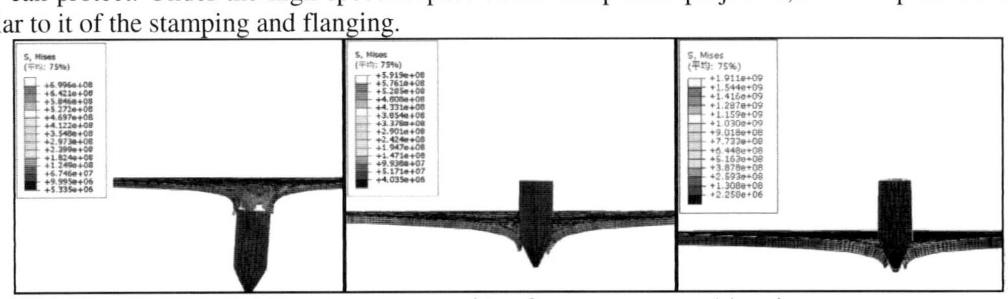

(a) t=2mm (b) t=3mm (c) t=4mm

Figure.7. Phenomena of sharp-nose projectile impacts on different thickness protective plate

4. Analysis on protection model of single-layer protective plate

Based on the above analysis, the relational mode of related parameter of the impact object damage-protective plate can be established by referring the blanking pressure and flanging force computational model[25] of the impact stamp die (refer to the formula 1). The coefficient is a correction coefficient based on the actual factors, e.g. change of impact speed, change of impact object mass, etc.

$$F = klt\tau, \begin{cases} l = 2\pi\sqrt{2rx - x^2} & x \le r, Ball-nose\ projectile \\ l = \dfrac{2\pi rx}{h} & x \le h, Sharp-nose\ projectile \\ l = 2\pi r & Flat-nose\ projectile \end{cases} \quad (1)$$

The shear-resisting strength τ depends on the type and shape of the thin plate material. For easy calculation, generally the estimated shear-resisting strength τ is taken as 80% of the strength limit σ_b of this material in the metal stamping technology, namely $\tau = 0.8\sigma_b$. For easy analysis, the material strength limit replaces the shear-resisting strength in this paper. The formula (1) is transformed to the formula (2).

IOP Conf. Series: Journal of Physics: Conf. Series **1016** (2018) 012019 doi:10.1088/1742-6596/1016/1/012019

$$F = 0.8klt\sigma_b \tag{2}$$

With growth of the protective plate thickness, the protection capabilities will gradually increase. If the protective plate with t_c thickness can just resist the impact object to the speed zero, namely the impact object is blocked inside the protective plate and does not pierce it, the thickness of this protective plate is called as the critical thickness. With momentum conservation transformation, the formula 2 is transformed to the formula (3).

$$F \cdot \Delta T = Klt\sigma_b \cdot \Delta T = mv_1 - mv_2 \tag{3}$$

The formula (3) can be further transformed to get the determination model of the critical thickness of the protective plate (refer to the equation 4), which is related to mass, speed and dimension of the impact object and the tensile strength of the protective plate.

$$t_c = \frac{mv_1}{Kl\sigma_b \cdot \Delta T} \tag{4}$$

In this formula:

F= Damage form of protective plate/N;

ΔT --time when the impact object switches from contact with the protective plate to separation from the protective plate (or be blocked inside the plate)/s;

m—mass of impact object/kg;

v1-- speed of impact object in case of contact with protective plate/ms/s;

v2--speed of impact object in case of separation from the protective plate (if it is blocked, the speed is 0)/m/s;

k--- correction coefficient for actual conditions;

K--- 0.8k;

l-- contact perimeter of the impact objects with protective plate in case of impact of impact object/mm;

t--- thickness of protective plate/mm;

x—moving distance since the elastic body contacts the protective plate/mm;

h—distance from the tip to the bottom/mm;

σ_b -- tensile strength of material/Mpa;

t_c -- critical thickness of protective plate.

5. Model validation

The impact object fails to pierce the Q235 protective plate in three impact for 3mm thick protective plate. The acting time with the whole protective plate is about 500μs. K value is calculated by using the final position of the elastic body and the contact perimeter of the protective plate in the model and the formula (3). For related values, refer to the table 1. The K values are approximate for three objects.

Table 1. Some related parameters when projectile impact speed drops to zero

Parameter name	deceleration time (s)	Contact perimeter (m)	σ_b (Pa)	m(kg)	V1 (m/s)	V2 (m/s)	K
Ball-nose projectile	550E-6	40.8E-3	410E6	42E-3	150	0	0.228
sharp-nose projectile	490E-6	47.1E-3	410E6	42E-3	150	0	0.222
flat-nose projectile	500E-6	47.1E-3	410E6	42E-3	150	-2.58	0.221

The above simulation results show that the reaction time of the impact object and protective plate is within hundreds of microseconds. If the mean K is 0.22 and the mean impact reaction time is 500μs, the formula (4) can be modified as formula (5):

$$t_c = 909 \cdot \frac{mv_1}{l\sigma_b} \tag{5}$$

In this formula, l is calculated by using the maximal contact perimeter of the impact object with the protective plate in case of impact. The formula (5) is used to compute critical thickness of the ball-nose projectile under different impact speed, which is compared with the thin plate thickness via impact simulation. The results show (refer to the figure 8) that all computed critical thicknesses resist penetration of the impact objects when $v_1 \le 340m/s$. When the critical thickness is smaller, most protective plates will be pierced, so the formula (5) is feasible in calculation of the critical thickness of the protective thin plate of the high-speed machine. According to figure 8, the impact velocity of formula (5) is limited to $v_1 \le 340m/s$.

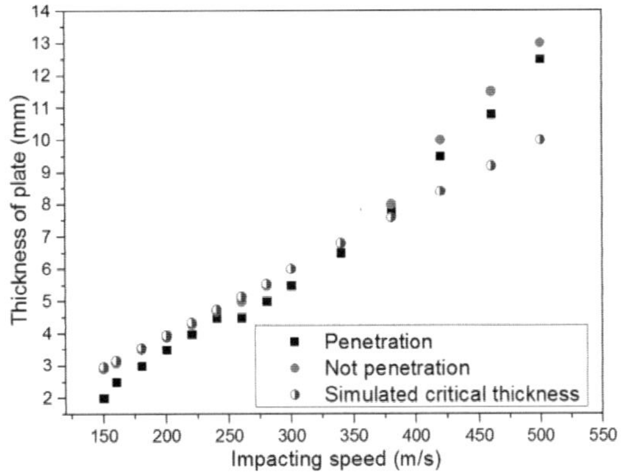

Figure.8. The resistance of different thickness protective plate for different impact speed

By analyzing 3mm thick Q235 protective plate in the experiments, the air cannon is used to fire the ball-nose projectile (42g) and the impact rate measured by the laser velometer is 168m/s. The experimental results are shown as the figure 8. The ball-nose projectile is blocked inside the protective plate and does not pierce the plate. The simulation results show that the protective plate is pierced. When the protective plate reaches 3.32 mm critical thickness calculated in the formula (5), the protective plate is not pierced. Simulation ignores factors such as elastic plastic deformation and heat loss in the true environment, so the simulated critical plate thickness is higher than the true critical plate thickness under equal impact conditions, namely the plate thicknesses calculated by the critical thickness calculation model (refer to the formula (5)) has certain safety tolerance.

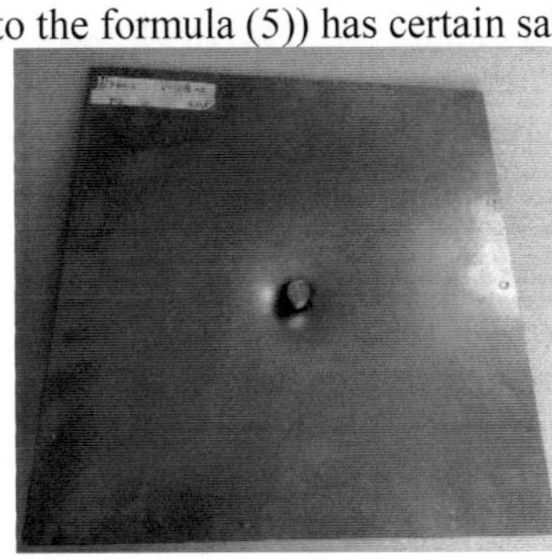

（a）The ball-nose projectile is blocked inside the plate

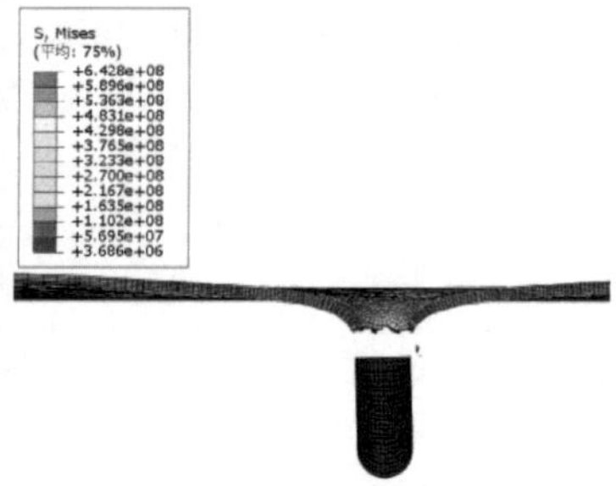

（b）Punching-like damage （c）Simulation results for piercing

Figure.9. High speed impact examination of ball-nose projectile

6. Conclusions

The impact resistance performance of the protective plate of the high-speed machine and simulates impact process is analyzed by using J-C model in this paper. The simulation results show that the protective plate damage complies with the fourth strength theory under the high-speed impact in the machine field. The shape-specific energy is the key factor to lead to material's yield damage. By combing the mechanical model in the plate stamping field, we deduce a new and easy-to-calculate critical thickness calculation model of the impact-resisting protective plate. The impact experiments show that the model results are reliable and have certain safety tolerance, so this model can be adopted as the theoretical support to identify the thickness of the protective plate of the high-speed machine.

References

[1] HouMingqiang, Gong Zizheng, Xu Kunbo, et al. Experimental study on hypervelo city impact characteristics of density-grade thin-plate, *Acta Physica Sinica*, 2014, **63** (2): 210-219.

[2] Li Yiyong, Shen Huairong, Li Zhi. Advances in hypervelocity impact dynamics and spacecraft protection research ,*Mechanics In Engineering*, 2009, **31** (2): 11-16.

[3] Zhang Yongqiang, Guan Gongshun, Zhang Wei, et al. Characteristics of debris cloud produced by normal impactof spherical projectile on thin plate plate, *Explosion And Shock Waves*, 2007, **27** (6): 546-552.

[4] Gai Fangfang, Liu Xianying, Liu Xun, et al. Numerical simulation on characteristics of debris clouds produced bycylindrical projectiles hypervelocity impact on thin plates, *Journal of Heilongjiang Institute of Science &Technology*, 2013, **23** (6): 600-604.

[5] Liu Xianying, Gai Fangfang, Li Zhiqiang et al. Characteristic Parameters of Debris Cloud Produced by HypervelocityImpact of Conical Projectiles on Spacecraft Plate, *Chinese Journal of High Pressure Physics*, 2016, **30** (3): 249-257.

[6] Chi RQ, Pang BJ, Guan GS, et al. Analysis of debris clouds produced by impact of aluminum spheres with aluminum thin plates. *International Journal of Impact Engineering*, 2008, **35** (12): 1465-1472.

[7] Rusinek A, Rodriguez-Martinez JA, Zaera R, et al. Experimental and numerical study on the perforation process of mild steel thin plates subjected to perpendicular impact by hemispherical projectiles. *International Journal of Impact Engineering*, 2009, **36** (4): 565-587.

[8] Gai Jingbo, Wang Shan, Tang ping. Damage of thin plate subjected to contact explosion loading, *Journal of Harbin Engineering University*, 2006, **27** (04): 523-525.

[9] Du Zhipeng, Li Xiaobin, Xia Lijuan, et al. Research on the Kinetic Armor-Piercing Effect of Semi-Armor-Piercing Missile Warhead, *Journal of Shanghai Jiaotong University*, 2006, **40**(10): 1794-1797.

[10] Dear JP, Lee H, Brown SA. Impact damage processes in composite thin plate and sandwich

honeycomb materials.*International Journal of Impact Engineering*, 2005, **32** (1-4): 130-154.

[11] Zhou Hao, Study on hypervelocity impact characteristics of corrugated sandwich structures.*Nanjing: Nanjing University of Science & Technology*,2016

[12] Guan Gongshun, Chen Liwen, Wang Shaoheng, et al. Experimental Investigation on Resist Capablility of Stainless Steel Mesh/Al Multi-Shock Plate by High-Velocity Impact. *Chinese Journal of High Pressure Physics*, 2012, 26 (2): 127-134.

[13] Yang Yang, Xu Fei, Kou Jiangfeng, et al. Impact resistance characteristics of C / Si C plate under hypervelocity load. *Journal of Vibration and Shock*, 2014, **33** (21): 189-192.

[14] Gotoh K, Sakamoto H, Ohbuchi Y, et al. Gun projectile collision and penetration problem in polycarbonate plate. *Rua Joao de Melo Abreu, Ponta Delgada, Azores, Portugal: Trans Tech Publications Ltd,* 2015: 257-260.

[15] Ude AU, Ariffin AK, Azhari CH. Impact damage characteristics in reinforced woven natural silk/epoxy composite face-thin plate and sandwich foam, coremat and honeycomb materials. *International Journal of Impact Engineering*, 2013, **58**(4): 31-38.

[16] Peng Jia. Research of the anti blast action of composite flexibleprotection structure,*Taiyuan: North University of China*,2015

[17] Gong Weiwei, Zhang Xiong. Material point method simulation of hypervelocity impact behavior of aluminum foam protective structure. *Mianyang, Sichuan, china*,2010:1

[18] Jia Bin, Ma Zhitao,Zhang Wei, et al. Numerical simulation investigation in ballistic limit of Whippleplate structure with Al-foam bumper, *Materials Science &Technology*, 2010, **18** (3): 368-372.

[19] Bi Qiang. Research on The Damage Characteristics of Ceramic Coating on Aluminum Bumper Plate Structure Under Hypervelocity Impact. *Harbin: Harbin Institute of Technology*,2011.

[20] Zhang XT, Jia GH, Huang H. An approach for constituting double/multi wall BLE by single wall BLE of spacecraft plate. *International Journal of Impact Engineering*, 2014, **69**(4): 114-121.

[21] Hayashida KB, Robinson JH. *Single wall penetration equations*. 1992.

[22] Elfer NC. *Structural Damage Prediction and Analysis for Hypervelocity Impacts: Handbook*. 1996.

[23] Johnson G R CWH. Fracture characteristics of three metals subjected to various strains, strain speeds, temperatures and pressures. *Engineering Fracture Mechanics*, 1985, **21** (85): 31-48.

[24] Pang Zirui. Experimental Study on Vitrified Bond CBN Wheel Based on Quick-point Grinding. *Shenyang: Northeastern University* ,2009

[25] Li Shuoben. Stamping technology: *China Machine Press*,1982

Acknowledgments

This work is supported by the Major Project of High-end CNC Machine Tools and Basic Manufacturing Equipment of China (2014ZX04011021), Project of Shaanxi Province(2016KTCQ01) .

IOP Conf. Series: Journal of Physics: Conf. Series **1016** (2018) 012020 doi:10.1088/1742-6596/1016/1/012020

Flowfield analysis of helicopter rotor in hover and forward flight based on CFD

Zhao Qinghe, Li Xiaodong

School of Energy and Power Engineering, Beihang University, Beijing 100191, China

Email: zhaoqinghe2001@163.com

Abstract. The helicopter rotor field is simulated in hover and forward flight based on Computational Fluid Dynamics(CFD). In hover case only one rotor is simulated with the periodic boundary condition in the rotational coordinate system and the grid is fixed. In the non-lift forward flight case, the total rotor is simulated in inertia coordinate system and the whole grid moves rigidly. The dual-time implicit scheme is applied to simulate the unsteady flowfield on the movement grids. The $k-\omega$ turbulence model is employed in order to capture the effects of turbulence. To verify the solver, the flowfield around the Caradonna-Tung rotor is computed. The comparison shows a good agreement between the numerical results and the experimental data.

1. Introduction

Helicopters have been used widely in the military, civil transportation, first aid domain and so on. To the present time the technology has already experienced great development. Modern helicopters request compact layout, high forward flight speed, manoeuvrability and low noise level, which lead to complex flowfield characteristics like following. Firstly, the flowfield include incompressible regions and transonic regions. Secondly, it is a typically unsteady flowfield, the airflow speed relative to the blades of the advancing side increase under the forward flight condition, some phenomenon of the transonic speed such as shock wave will appear. On the other side, the relative speed decrease with the flow angle is comparatively high usually lead to dynamic stall. Thirdly, there are the tip vortex which make a strong impact on the aerodynamic characteristic of flight vehicle.

After rapid development, computational fluid dynamics(CFD) has received practical application in aviation, as a result of the mentioned complexity above, its application to rotors is much less than fixed wings. [1-6] The grid system is an important issue in rotorcraft numerical simulation. The Chimera approach was used to solver the Reynolds-averaged Navier-Stokes equations on a multi-block mesh. The blade grids are embedded in a background mesh. Along the Chimera boundaries, a tri-linear interpolation method is used to couple the solution on the blade-fixed and background grids. Another approach is using sliding planes method. It couple a part of the mesh rotating with the rotor and a stationary lower part around an approximate helicopter fuselage. Besides, due to the convenience in the grid generation, the unstructured grid is also used wildly.

In the present work, multi-block structured meshes are used to model the non-lift forward flight with a rigid mesh motion. This approach was designed to overcome the complexity of the Chimera grid. The structure of the paper is as follows. Section 2 presents spatial discretization and temporal integration method in the rotational coordinate system and in inertia coordinate system. Section 3 gives the validation results for both hovering and forward flight rotor. Some conclusions are finally drawn in Section 4.

2. Numerical Solution Procedures

Assuming that the wake shed from the rotor is steady, the flow around a hovering rotor can be treated as a steady-state problem. For forward-flight simulations, the flow field is asymmetry and unsteady. So it introduces the simulation process separately.

2.1. Methodologies for steady flows

For hover cases, due to the symmetry of the flow in rotational coordinate system, the computational domain can be replaced by a domain around one blade. The periodicity boundary condition is used to taken into account the influence of other blades. Furthermore, the periodicity boundary condition can be used to reduce the computational expense. The government equation is given below

$$\frac{d}{dt}\int_{V(t)} \vec{w}dV + \int_{\partial V(t)} (\vec{F}(\vec{w}) - \vec{F}_v(\vec{w}))\vec{n}dS = \vec{S} \tag{1}$$

Where $\vec{w}$ is the conserved variables. $\vec{F}$ and $\vec{F}_v$ are the inviscid and viscous fluxes, respectively. There is a source terms in the non-inertial frame of reference. It means the influence of the centripetal and the Coriolis force. For an hovering rotor rotating around the y-axis with the angular velocity $\vec{\omega} = [0, \Omega, 0]$, the source term $\vec{S}$ can be expressed by equation 2.

$$\vec{w} = \begin{bmatrix} \rho \\ \rho u \\ \rho v \\ \rho w \\ \rho E \end{bmatrix}, \quad \vec{F} = \begin{bmatrix} \rho(q - q_\omega) \\ \rho u(q - q_\omega) + p i_x \\ \rho v(q - q_\omega) + p i_y \\ \rho w(q - q_\omega) + p i_z \\ \rho h(q - q_\omega) + p q_\omega \end{bmatrix}, \quad \vec{S} = \begin{bmatrix} 0 \\ -\rho\Omega w \\ 0 \\ \rho\Omega u \\ 0 \end{bmatrix} \tag{2}$$

Where q_ω is the grid velocity, it produce a grid flux. The finite volume discretization with central differencing for the flux approximation leads to 2 order scheme. The implicit Lower-Upper Symmetric Gauss Seidel operator(LUSGS) is applied to obtain the update solution. Besides, the mesh around the rotor does not rotate about the rotational axis.

2.2. Methodologies for unsteady flows

For forward flight, the flowfield should be simulated in unsteady condition. The dual time stepping approach is introduced, and the system can be solved by an iteration through pseudo time to the steady state, as given by

$$\frac{3V_{i,j,k}^{n+1} w_{i,j,k}^{n+1} - 4V_{i,j,k}^{n} w_{i,j,k}^{n} + V_{i,j,k}^{n-1} w_{i,j,k}^{n-1}}{2\Delta t} + R_{i,j,k}(w_{i,j,k}^{n+1}) = 0 \tag{3}$$

$$V_{i,j,k}^{n+1} \frac{w_{i,j,k}^{n+1,m+1} - w_{i,j,k}^{n+1,m}}{\Delta\tau} + \frac{3V_{i,j,k}^{n+1} w_{i,j,k}^{n+1,m} - 4V_{i,j,k}^{n} w_{i,j,k}^{n} + V_{i,j,k}^{n-1} w_{i,j,k}^{n-1}}{2\Delta t} + R_{i,j,k}(w_{i,j,k}^{n+1,m}) = 0 \tag{4}$$

$$V_{i,j,k}^{n+1}\left(\frac{1}{\Delta\tau} + \frac{3}{2\Delta t} + A + B + C\right)\Delta w_{i,j,k} = -R_{i,j,k}(w_{i,j,k}^{n+1,m}) - \frac{3V_{i,j,k}^{n+1} w_{i,j,k}^{n+1,m} - 4V_{i,j,k}^{n} w_{i,j,k}^{n} + V_{i,j,k}^{n-1} w_{i,j,k}^{n-1}}{2\Delta t} \tag{5}$$

Where Δt is the physical time step, $\Delta\tau$ is the pseudo time step, $\Delta w_{i,j,k} = w_{i,j,k}^{n+1,m+1} - w_{i,j,k}^{n+1,m}$, A, B, C is the Jacobi metrics of the inviscid flux. The accelerate convergence methods such as local time step can be used in the pseudo time cycles.

2.3. The grid topology and the boundary conditions

The grid generation procedures is an important issue for simulation, especially for complex flowfield and the multiple bodies in relative motion. Because the grid quality effects the accuracy and convergence rate of numerical solutions. Furthermore, in hover simulations, the periodicity boundary are required to form a perfect match at the boundary to avoid data interpolation errors. Figure 1 shows the topology, and the enlarged figure is given in Figure 2. The rotor centre of rotation is laid at the y-axis, the rotor blade is located on the x-axis and the quarter chord point is laid on at z-axis. The blade leading edge points at the negative x-axis direction.

The computational domain of the rotor is divided into 22 Multi-blocks. An O type blocking topology around the blade is used. The accurate viscous case computation can be performed in this type of mesh topology. There are 161 points along the airfoil surface with concentration near leading and trailing edges. The first cell size is 10^{-5} of the blade chord length normal to the surface and the cell ratio follows an exponential law. The rotor hub was modelled as a simple straight cylinder.

The computational domain is surrounded by 4 types boundary conditions: viscous wall, rotational periodicity, characteristic type boundary and the hub conditions. To maintain second-order accuracy at the boundaries, the domain is surrounded by two layers of ghost grids. Take the viscous wall boundary condition as an example

$$u_{ghost} = 2u_{wall} - u_1,\ v_{ghost} = 2v_{wall} - v_1,\ w_{ghost} = 2w_{wall} - w_1 \quad (6)$$

where subscript ghost denotes the ghost grid, subscript wall denotes the value on the wall, it equal the wall movement velocity, in particular it equal 0 when the wall keep static. The subscript 1 denotes the velocity of the first level grid.

The hub conditions are applied when performing single-blade simulations, in hover and forward flight conditions. Characteristic type boundary conditions are applied at the far-field. They are based on free-stream values when the far field boundary is located relatively far from the rotor.

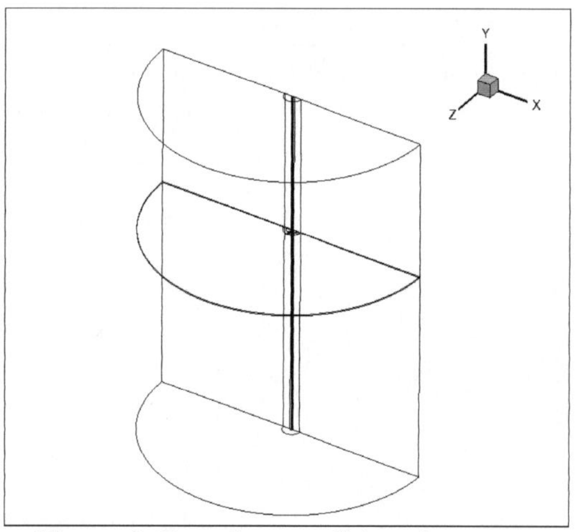

Figure 1. Boundary conditions for a rotor

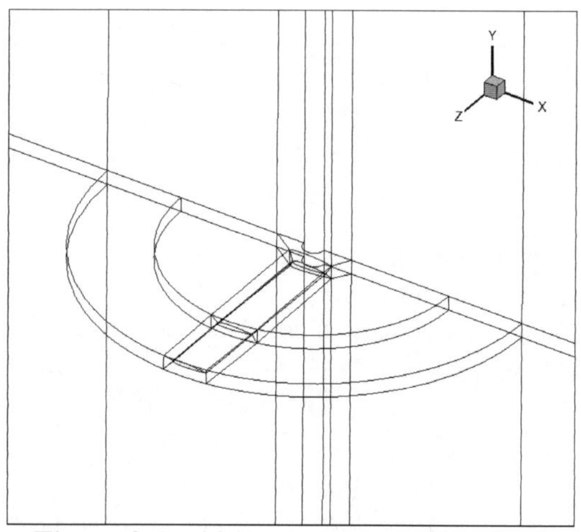

Figure 2. Multi-block topology for a rotor

2.4. Mesh movement

The mesh movement should be used under the forward-flight condition. There is no change of the relative position of points in the same grids if the rotor moves rigidly. Assuming it rotate around the y-axis, the transformation matrices are introduced for rotation. And the grid coordinate and grid velocity can be calculated directly from formula 7. Where 0 denotes the original grid position and n denotes the updated grid position.

IOP Conf. Series: Journal of Physics: Conf. Series **1016** (2018) 012020 doi:10.1088/1742-6596/1016/1/012020

$$\begin{bmatrix} x \\ y \\ z \end{bmatrix}^n = \begin{bmatrix} \cos\omega t & 0 & \sin\omega t \\ 0 & 1 & 0 \\ -\sin\omega t & 0 & \cos\omega t \end{bmatrix} \begin{bmatrix} x \\ y \\ z \end{bmatrix}^0 \quad \begin{bmatrix} u \\ v \\ w \end{bmatrix}^n = -\omega \begin{bmatrix} \sin\omega t & 0 & -\cos\omega t \\ 0 & 0 & 0 \\ \cos\omega t & 0 & \sin\omega t \end{bmatrix} \begin{bmatrix} x \\ y \\ z \end{bmatrix}^0 \tag{7}$$

There is no grid volume changing, so the geometric conservation law(GCL) is also kept.

3. Validation of the framework

Validation is now presented for the hover formulation and the forward flight method. A summary of cases is given in Table 1. The Mach number and the Reynolds number are based on rotor tip velocity. Where μ is the advance ratio, which means the ratio of the free flow velocity and the tip velocity. All the test case have been performed with the $k-\omega$ turbulence model.

Table 1. Simulation conditions of the helicopter rotors in hover and forward flight

	Ma	*Re*	μ
Hover	23.56	2.27×10^6	0
Forward flight	27.9	3.58×10^6	0.2

3.1. *In hover*

An experimental and analytical study of a model helicopter in hover had been carried out by Caradonna and Tung [7]. The blade had a NACA0012 profile and were untwisted and untapered, which had a diameter of 2.286 meter, and a chord length of 0.191m, according to the aspect ratio was 6. In the Caradonna-Tung experiment, the surface pressure distribution was measured at five rotor blade sections (r/R=0.50,0.68,0.80,0.89,0.96). The Cp is calculated by formula 8, where ω is the rotate velocity and r is the span distance.

$$Cp = \frac{p - p_0}{0.5\rho(\omega r)^2} \tag{8}$$

The CFD results for the pressure coefficient distribution at four different sections of the blade were compared to the experimental data in Figure 3. The comparison shows a good agreement between the numerical results and the experimental data.

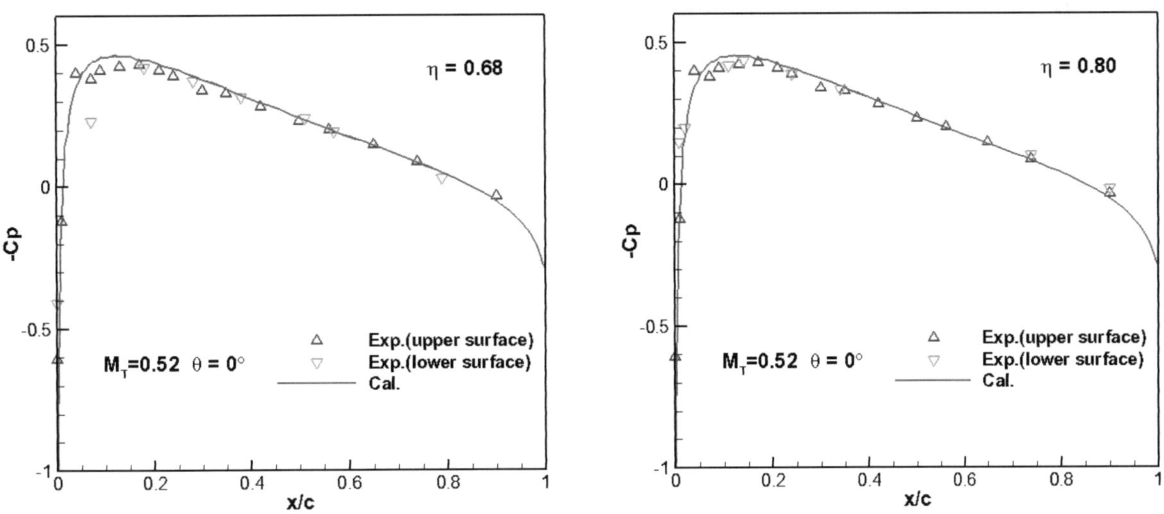

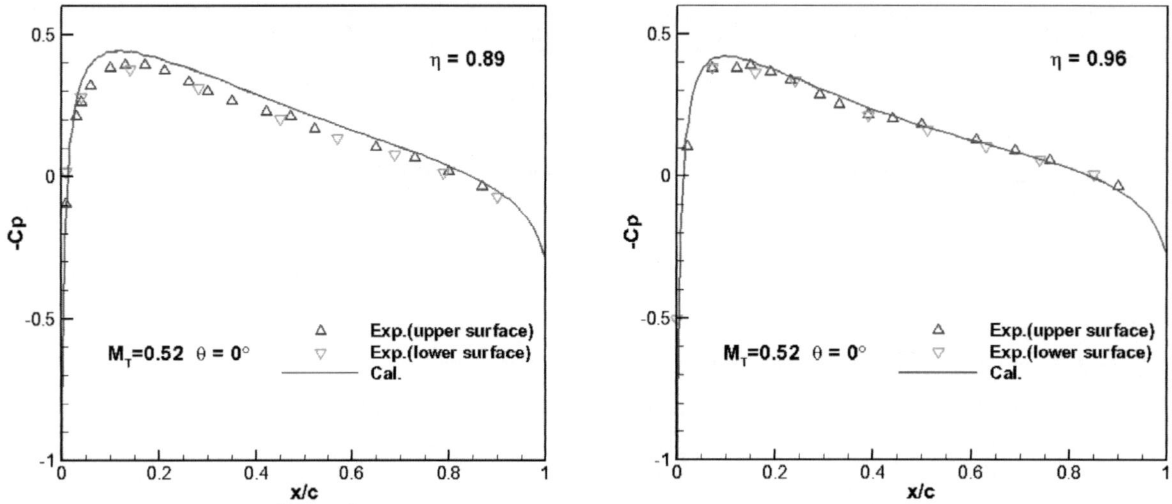

Figure 3. Pressure distribution comparison on the hovering rotor

3.2. Non-lifting forward flight

The case is for a non-lifting rotor which has two blades and the airfoil is NACA0012. The rotor aspect ratio was 7. The experimental data [8] is available for the surface pressure on the advancing side. The grid moves rigidly with the blade surface. Therefore, the rotation is applied to the whole grid. The hover tip Mach number is 0.8 and advanced ratio is 0.2. There is a strong transonic effects on the advancing blade. The azimuth angle increment of 1° is used for the computation. The surface pressure distributions according to six azimuth angles $\varphi = 30^\circ, 60^\circ, 90^\circ, 120^\circ, 150^\circ, 180^\circ$ are shown in Figure 4. The wind flows from left to right, and the rotor rotate in anticlockwise direction. A strong shock wave is seen on the advancing side of the rotor blade.

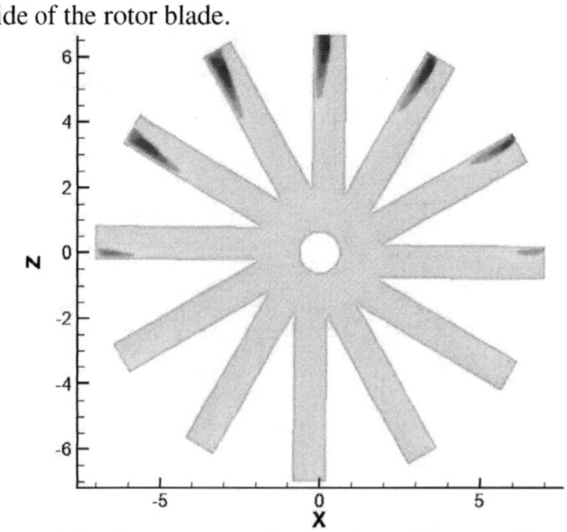

Figure 4. Surface pressure distribution of forward flight rotor

Figure 5 shows the surface pressure distributions at 89% span for six azimuth angles. The Cp is calculated by formula 9, where V_∞ is the free stream velocity and φ is the azimuth angle. There are some difference between the numerical results and the experimental data. The flow is highly unsteady and the shock is moving rapidly downstream.

$$Cp = \frac{p - p_0}{0.5\rho(\omega r + V_\infty \cdot \sin\varphi)^2} \tag{9}$$

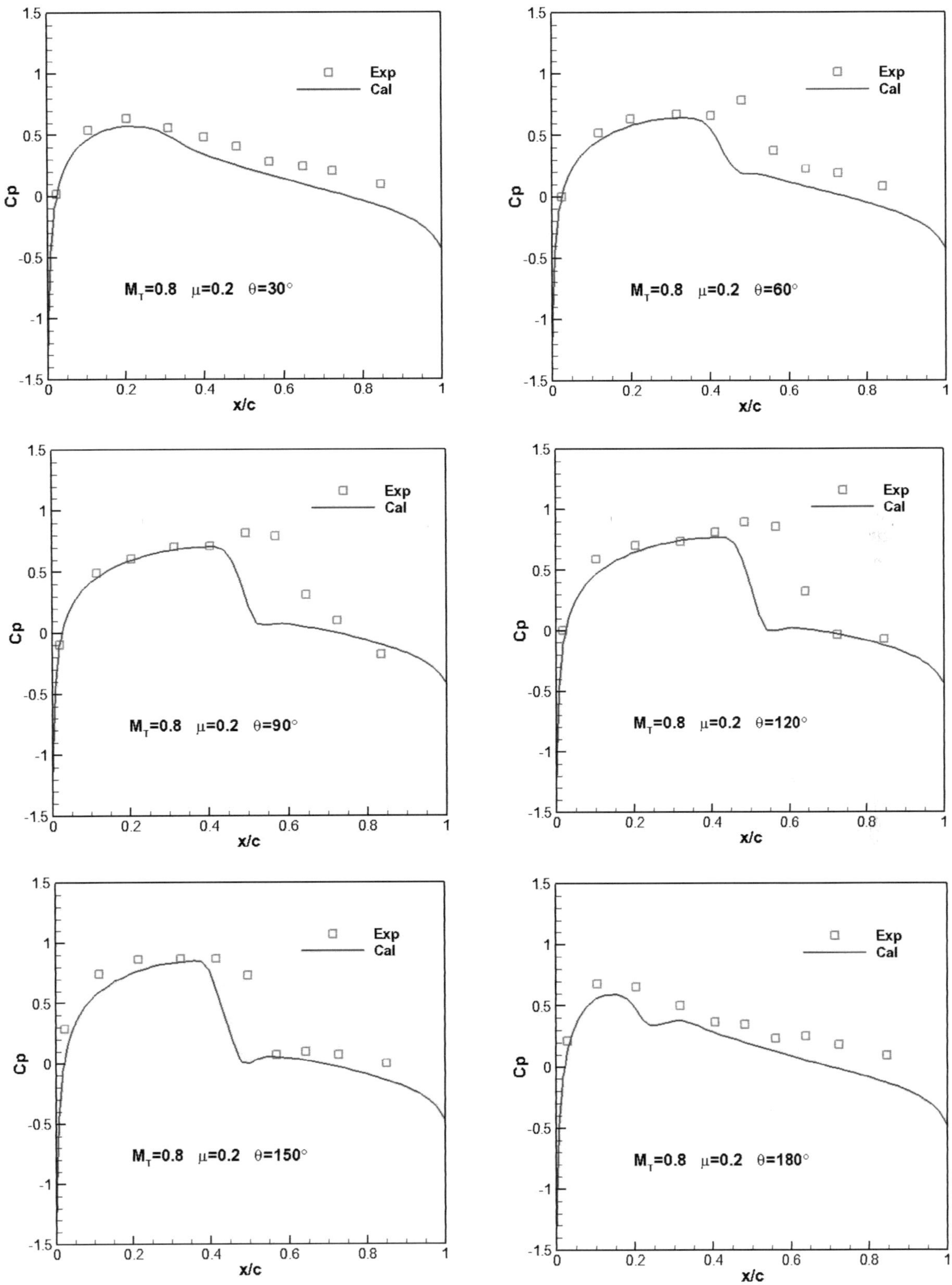

Figure 5. Pressure distribution comparison on the forward flight

IOP Conf. Series: Journal of Physics: Conf. Series **1016** (2018) 012020 doi:10.1088/1742-6596/1016/1/012020

4. Conclusions

The helicopter rotor field is simulated in hover and forward flight. In hover case the single rotor field is simulated in the rotational system and the grid is fixed. In the non-lift forward flight case, the total rotor is simulated in inertia system, the whole grid move rigidly, The finite-volume solver adopts a dual-time implicit scheme on movement grids. The $k-\omega$ turbulence model is employed in order to capture the effects of turbulence. The unsteady flowfield of the Caradonna-Tung rotor has been investigated. The surface pressure distribution compared well with experimental data. Forward flight prediction capabilities have been validated through the Caradonna-Tung rotor. Some conclusions are drawn below:

a) In the rotational frame the hover flowfield can be transformed from the unsteady problem to a steady one. The accuracy can be kept and the simulation time decrease greatly.
b) In forward flight condition the rotor flowfield is unsteady. It can be simulated by dual-time marching with the moving mesh method.

References

[1] Hariharan N Egolf A and Sankar L 2014 Simulation of Rotor in Hover: Current state and challenges *AIAA paper* 2014-0041
[2] Srinivasan G R and Baeder J D 1992 Flowfield of lifting rotor in hover: A Navier-Stokes simulation *AIAA Journal* Vol 30 No 10
[3] Jacques S Klansdieter P and Michel C 2001 Numerical simulation of flows around helicopters at DLR and ONERA *Aerosp Sci Technol* 5:35-53
[4] Pomin H and Wagner S 2002 Navier-Stokes analysis of isolated rotor flow in helicopter hover flight *Journal of Aircraft* 39(5) 813-821
[5] Steijl R Barakos G and Badcock K 2006 A framework for CFD analysis of helicopter rotors in hover and forward flight International *Journal for numerical methods in fluids* 51:819-847
[6] Renzoli P D'Alascio A Kroll N Peshkin D and H.L.Hounjet M 2000 EROS-a common European Euler code for the analysis of the helicopter rotor flowfield *Progress in Aerospace Sciences* 36:437-485
[7] Caradonna F X and Tung C 1981 Experimental and analytical studies of a model helicopter rotor in hover *Vertica* 5(1):149-161
[8] Caradonna F X Laub G H and Tung C 1984 An experimental investigation of the parallel blade-vortex interaction *NASA/TM* 86005

Probability density function evolution of power systems subject to stochastic variation of renewable energy

J Q Wei[1], Y C Cong[1] and M Q Xiao[1]

[1]School of Mathematical and Physical Science,North China Electric Power University Beijing 102206, P.R. China

Email: weijunqiang@ncepu.edu.cn

Abstract. As renewable energies are increasingly integrated into power systems, there is increasing interest in stochastic analysis of power systems.Better techniques should be developed to account for the uncertainty caused by penetration of renewables and consequently analyse its impacts on stochastic stability of power systems. In this paper, the Stochastic Differential Equations (SDEs) are used to represent the evolutionary behaviour of the power systems. The stationary Probability Density Function (PDF) solution to SDEs modelling power systems excited by Gaussian white noise is analysed. Subjected to such random excitation, the Joint Probability Density Function (JPDF) solution to the phase angle and angular velocity is governed by the generalized Fokker-Planck-Kolmogorov (FPK) equation. To solve this equation, the numerical method is adopted. Special measure is taken such that the generalized FPK equation is satisfied in the average sense of integration with the assumed PDF. Both weak and strong intensities of the stochastic excitations are considered in a single machine infinite bus power system. The numerical analysis has the same result as the one given by the Monte Carlo simulation. Potential studies on stochastic behaviour of multi-machine power systems with random excitations are discussed at the end.

1. Introduction

There are many stochastic factors in the operation of power system, such as generator outages, short-circuits or generation and load level fluctuations.In recent years, the stochastic factors in power system haveshownhigh complexity and volatilitywith the expansion of renewable energy resources and wide-spread adoption of plug-in electric vehicles.These stochastic factors can cause unbalanced and stochasticpower fluctuation in the operation of power systems, which brings great threat to stability of power systems, thus researching the stability of power system has a great realistic meaning[1-2]. The traditional methods of Differential Algebraic Equations (DAEs) cannot fully reflect the impact of stochastic excitations on power system; hence SDEs are employed to describe the stochastic characteristics of power system to better describe the stochastic factors[3-4]. The theory of SDEs has been applied widely in many fields, but it is still in the process of exploration on power system analysis. J.Y. Zhang constructed astochasticmodel for single machine connected to an infinite bus (SMIB) system under Gaussian type random excitation with the application ofSDEs, then the operating trajectorieswereobtained through numerical method, finallythe mean stability and the mean square stability were proved theoretically[1]. A stochastic model of wind turbine based on SDEs was established in [5], a criterion for stability under random excitationwas put forward through Monte Carlo simulation subsequently.Because of the theoreticalcomplexity of SDEs and the high dimensionality and strong nonlinearity of power system, it is difficult to obtain the analytic solution.

Most of the existing methods to obtain the solution generally refer to the numerical approach or Monte Carlo method.Andit is generally believed that the numerical solution is a feasible tool tostudy the trajectoryof stochastic modelof power systems[6].Among the numeric methods, Euler-Maruyama method and Milstein's method are commonly used.A systematic introduction to the numerical simulation of SDEs can be found in [7-8].

Another effective method to study SDEs model is focus on the evolution of the probability density function (PDF) [9]. It has been proved that the solution of ItôSDEsis a continuousMarkov and diffusion process[10-11]. According to the theory of stochastic dynamics, if a diffusion proceed the solution of SDEs, then the PDFsatisfies a Fokker-Planck-Kolmogorov (FPK) equation[6][9].And the FPK equation can be used to determine the PDFandthe steady probability density of the stochastic process, then by observing the shape of stationary PDF as parameter changes,stochastic bifurcation behaviour of the power systems can be explored [12].However, studiesfocusing on this method is relatively rare. Reference[13] established the FPK equation of reliability function of SMIB system operation reliability, and thenanalysed the first passage problem.A dynamic model for SMIB systemwith stochastic excitation was established in [12], then through simplified the FPK equation properly, the approximate analytic solution of the stationary probability density was obtained, but numerical solution of theFPKequation was not given.It is difficult to obtain analytic solution of theFPKequation except special case, so numerical methodsare necessary to obtain the numerical solution. Z.L. Huang proposed an algorithm for solving the numerical solution of the high dimensional stationary FPK equation [14]. And a systematic introduction to the numerical solution of FPKequation can be found in [15].

The remainder of this paper is arranged as follows. In the next section, the mathematical formulation of new energy power systems based on SDEs are introduced. In the third section, the stochastic probability density function and its evolution arediscussed.Case studies for SMIB system are done in the fourth section. And in the last section, the conclusions on the stability problems caused by stochastic excitations of renewable energies and potential studies are discussed.

2. Mathematical Formulation of Power System with Stochastic Excitations by Renewables

2.1. Itô Stochastic Differential Equation

The general Itô stochastic differential equation is as follows [10-11]:

$$\begin{cases} \mathrm{d}\boldsymbol{x}(t) = \boldsymbol{a}(\boldsymbol{x}(t), \boldsymbol{t})\mathrm{d}t + \boldsymbol{G}(\boldsymbol{x}(t), \boldsymbol{t})\mathrm{d}\boldsymbol{B}(\boldsymbol{t}) \\ \boldsymbol{x}(t_0) = \boldsymbol{x}_0 \end{cases} \tag{1}$$

where$\boldsymbol{x}(\boldsymbol{t})$ and $\boldsymbol{a}$ are $n \times 1$ vectors and $\boldsymbol{x}(\boldsymbol{t})$ is a stochastic process, $\boldsymbol{G}$ is an $n \times m$matrix, $\boldsymbol{B}(\boldsymbol{t})$ is a $m \times 1$ vector of Wiener processes with covariance matrix $\boldsymbol{V}$. $\mathrm{d}\boldsymbol{B}(\boldsymbol{t})$is the Gaussian noise excitation corresponding to the Wiener process, which is also indicated as $\boldsymbol{W}(\boldsymbol{t})$.

2.2. SMIB Power Systems with Stochastic Excitation

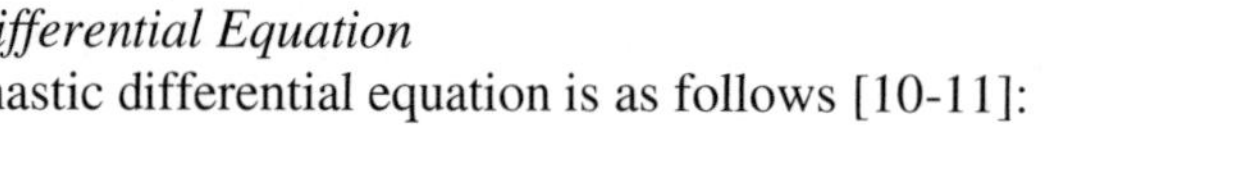

Figure 1. Single Machine Infinite Bus (SMIB) system.

As the renewables connected to the power system shown in Figure 1, the stochastic model of SMIB system can be constructed as follows under proper assumptions [16]:

$$M\ddot{\delta} + D\dot{\delta} = P_{\mathrm{m}} - P_{\mathrm{e}} + \sigma W(t) \tag{2}$$

whereδ is the relative generator angle to the infinite bus. $\dot{\delta}$ and$\ddot{\delta}$are derivative of first order and second order of δ, and $\dot{\delta}$is the generator velocity of the system.$P_{\mathrm{e}}$is the electrical power and $P_{\mathrm{e}} = P_{\max}\sin\delta$, where $P_{\max}$ is a constant. The constants M, D and P_{m} are the machine inertia constant, damping coefficient and the mechanical power, respectively. $W(t)$is additive white noise which indicates random excitation of renewable energies, such as power fluctuation of wind turbines and solar panels, and/or load swing of plug-in electric vehicles. $\sigma$is the magnitude of $W(t)$, which indicates intensity of the random excitation.

Using$\omega$to represent$\dot{\delta}$, and then equation (2) can be equivalently written as follows:

$$\begin{cases} \mathrm{d}\delta = \omega \mathrm{d}t \\ \mathrm{d}\omega = \left(\tilde{P}_\mathrm{m} - \tilde{P}_\mathrm{e} - \tilde{D}\omega\right)\mathrm{d}t + +\tilde{\sigma}W(t) \end{cases} \tag{3}$$

where $\tilde{P}_\mathrm{e} = \frac{P_\mathrm{e}}{M}; \tilde{D} = \frac{D}{M}; \tilde{P}_\mathrm{m} = \frac{P_\mathrm{m}}{M}; \tilde{\sigma} = \frac{\sigma}{M}$.

The vector form of equation(3) is:

$$\mathrm{d}\boldsymbol{x} = f(\boldsymbol{x})\mathrm{d}t + P_\mathrm{L} \tag{4}$$

where $\boldsymbol{x} = (\delta, \omega)^T$, $P_\mathrm{L} = [0, \tilde{\sigma}W(t)]^T$, and $f(\boldsymbol{x}) = [f_1(\boldsymbol{x}), f_2(\boldsymbol{x})]^T = (\omega, \tilde{P}_\mathrm{m} - \tilde{P}_\mathrm{e} - \tilde{D}\omega)^T$.

For the power system described by equation (3), the stability of the system under small random excitation is of great interest. From the dynamic viewpoint, the power system can always maintain an equilibrium position (i.e., the equilibrium point of the system) to operate smoothly without perturbations, and the stability of the power system under small excitations is the ability of the system to continue to be in the equilibrium position or to achieve a new equilibrium state after impacted by random excitations. And the stability of the power system can be reflected by the stationary PDF.

3. Probability Density Function (PDF) and its Evolution

3.1. Fokker-Planck-Kolmogorov (FPK)Equation

In general, conditional transition probability density $p(\boldsymbol{x}, t|\boldsymbol{x}_0, t_0)$ indicates the probability density of the state at time t under the condition of initial state $(\boldsymbol{x}_0, t_0)$. As mentioned earlier, $p(\boldsymbol{x}, t|\boldsymbol{x}_0, t_0)$ satisfies following FPK equation [17]:

$$\frac{\partial p(x,t|x_0,t_0)}{\partial t} = -\sum_{j=1}^{n} \frac{\partial}{\partial \mathrm{x_j}} [\boldsymbol{a}_i(\boldsymbol{x}, t)p(\boldsymbol{x}, t|\boldsymbol{x}_0, t_0)] + \sum_{i,j=1}^{n} \frac{1}{2} \frac{\partial}{\partial \mathrm{x_i}\, \partial \mathrm{x_j}} [\boldsymbol{b}_{ij}(\boldsymbol{x}, t)p(\boldsymbol{x}, t|\boldsymbol{x}_0, t_0)] \tag{5}$$

where $\boldsymbol{b}_\mathrm{ij}(\boldsymbol{x}, \mathrm{t}) = [\boldsymbol{GVG}]^T$.

Solving equation (5) requires initial condition, boundary condition and normalization condition. Here the initial condition can be given by $p(\boldsymbol{x}, t|\boldsymbol{x}_0) = \prod_{\mathrm{j}=1}^{n} \delta(x_\mathrm{j} - x_{0\mathrm{j}})$as $t \to t_0$, where $\boldsymbol{x_0}$ is the initial state and$\delta$is the Dirac delta function. For power system, the boundaries can be assumed to natural boundary conditions [16], which indicated that $p(\boldsymbol{x}, t|\boldsymbol{x}_0, t_0) = 0, \boldsymbol{x} \in \boldsymbol{\Gamma}$, where $\boldsymbol{\Gamma}$ is the boundary of security domain. A normalization condition is also need to be satisfied:

$$\int p(\boldsymbol{x}, t)d\boldsymbol{x} = 1 \tag{6}$$

For the most part of engineering problems, the stationary solutionof FPKequation,which means thebehaviour of the system dynamics as the states approaches steadystate, is of greatest interest. If the stationary solution exists as $t \to \infty$, then $p(\boldsymbol{x}|\boldsymbol{x}_0)$is indicated the stationary PDF and $p(\boldsymbol{x}|\boldsymbol{x}_0) = \lim_{\mathrm{t}\to\infty} p(\boldsymbol{x}, t|\boldsymbol{x}_0, t_0)$. Under this steady-state condition, $\partial p/\partial t = 0$, then equation (5) is called stationary FPK equation.

3.2. Numerical Methods for FPK Equation

As mentioned earlier, the analytic solution of FPK equation only can be obtainedunder restrictive conditions, thus numerical methods are necessary to calculateapproximatesolutions.Mathematically, FPK equation (5) is a parabolic Partial Differential Equation (PDE) with variable coefficients; therefore computational methods for PDE are adopted, such as path integration method, finite difference method, finite element method and Monte Carlo method, etc[15]. Herein, finite difference method was used, andsecond-order central difference schemes were chosento get the discrete equation [14].Then linear algebraic equations can be obtained from the discretized equation as follows:

$$AP = 0 \tag{7}$$

where A is coefficient matrix and P is an $\bar{M} \times 1$vectorconnected by stationaryPDFon the nodes that need to be calculated and $\bar{M}$ is the number of the nodes of discretization.

Finally, the numerical solution of equation (5) can be obtained by solving equations (7). To solve the linearequations, we adopted the numerical solution combined the successive-over-relaxation (SOR) method.Corresponding iterative scheme of SOR is as follows[18]:

$$x_{\mathrm{i}}^{(\mathrm{n}+1)} = x_{\mathrm{i}}^{(\mathrm{n})} + \alpha(-\sum_{\mathrm{j}=1}^{\mathrm{i}-1} A_{\mathrm{ij}} x_{\mathrm{j}}^{(\mathrm{n}+1)} - \sum_{\mathrm{j}=1}^{\bar{M}} A_{\mathrm{ij}} x_{\mathrm{j}}^{(\mathrm{n})})/A_{\mathrm{ii}} \tag{8}$$

where α is the relaxation factor that takes value on interval$(0, 2)$,generally the appropriate value can be given by$\alpha = 2[1 + e(A)]^{-1/2}$, where $e(A)$ is the condition number of$A$. And note that the results should be normalized by equation (6).

4. Case Studies for SMIB System

4.1. The Stationary FPK Equation of SMIB System

For stability analysis of power system, the operation of the steady-state after being disturbed is focused. According toequation (3) and (5), thestationary FPK equation that SMIB systemsatisfied can be given as follows:

$$\frac{\sigma^2}{2}\frac{\partial^2 p(\boldsymbol{x})}{\partial \omega^2} - \frac{\partial[f_1(\boldsymbol{x})p(\boldsymbol{x})]}{\partial \delta} - \frac{\partial[f_2(\boldsymbol{x})p(\boldsymbol{x})]}{\partial \omega} = 0 \tag{9}$$

And the exact stationary solutionof equation (9) is [17]:

$$p(\boldsymbol{x}) = c\exp\left[-\frac{2\tilde{D}}{\sigma^2} \cdot \left(\frac{\omega^2}{2} - \tilde{P}_{\max}\cos\delta - \tilde{P}_{\mathrm{m}}\delta\right)\right] \tag{10}$$

where c is the normalized constant.

4.2. Numerical Simulation of Stationary FPK Equation for SMIB System

For the stationaryFPK equation (9) of SMIB system with the randomexcitationof renewable energies, numerical simulation was carried out according to the numerical method above. The system parameters are$\tilde{D} = 0.41; \tilde{P}_{\max} = 1; \tilde{P}_{\mathrm{m}} = 0.5; \boldsymbol{x}_0 = (0,1.5)^T$and the equilibrium pointwas calculated as $\boldsymbol{x}_e = (0.524,0)^T$.

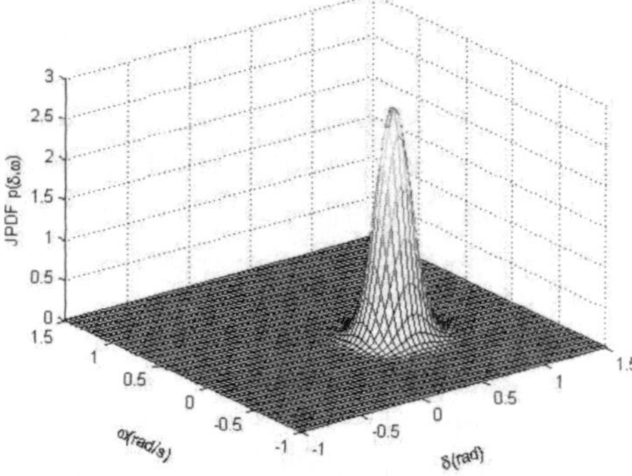

Figure 2. Theoretical solutionof stationary PDF

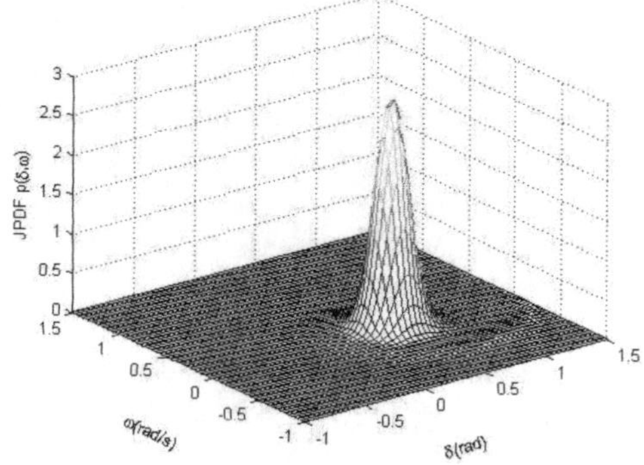

Figure 3. The numerical solutionof stationary PDF

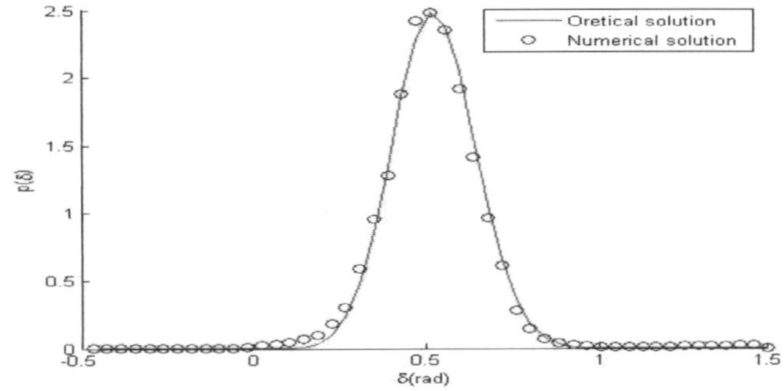

Figure 4. Comparison of theoretical and numerical results

Figure 2 shows the exact stationary solution with $\sigma = 0.1$, the numerical results shown in Figure 3and Figure 4 gives a comparison of the exactstationary solution and numerical solution of marginal probability.The results illustrate the effectiveness of the numerical method. As is shown in the figures, the stationary PDF is clustered in a peak around theequilibrium point of the system. It is shown that the system of equation (3) can maintain the stability under the stochastic perturbations with $\sigma = 0.1$.

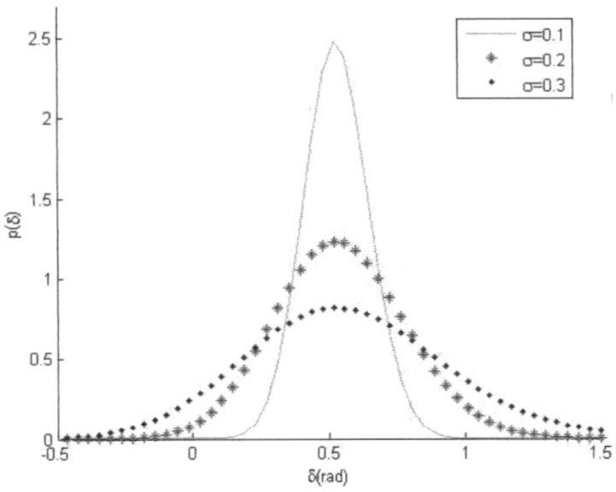

Figure 5. Marginal probabilitywith various intensities of stochastic excitation

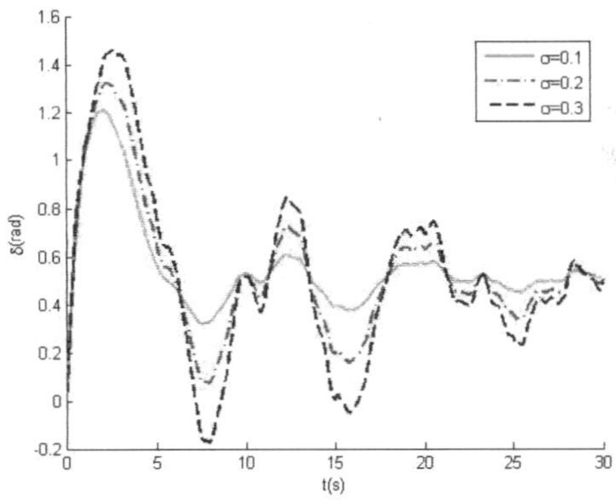

Figure 6. Trajectories with various intensities of stochastic excitation

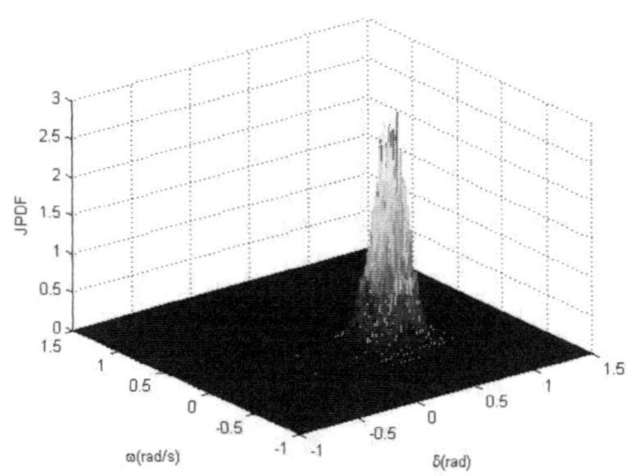

Figure 7. Simulation of stationary PDF by Monte Carlo method

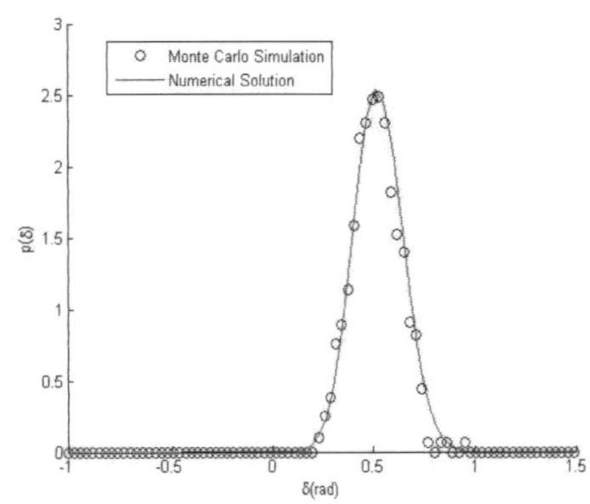

Figure 8. Comparison of numerical solution and Monte Carlo simulation of marginal probability

Figure 5 shows impacts of the evolution on Marginal probability$p(\delta)$ under different intensities of random excitation. It can be concluded that stationary PDF are concentrated near the equilibrium point and the concentration trend weakensas the intensity gettingstronger.The results indicate that when the power system is subjected to excitations from renewable energies or other factors, the greater the intensity of disturbances is, the lower the stability of the system is.Euler-Maruyama method was used to obtain the operatingtrajectories of the system under different intensities of stochastic excitation, which is shown in Figure 6. It can be seen from the figure that each trajectory has a vibration due to the existence of disturbances, and the amplitude of vibration increases when the intensity gets stronger, which is consistent with the result in Figure 5.

In order to verify the correctness of the conclusions, Monte Carlo method is used to simulate the system of equation (4) 1000 times. Figure 7 and 8 show the result of Monte Carlo method.Figure 7 is the result of stationary PDF, which is corresponding to the solution obtained by FPK equation. And from the comparison in Figure 8, the results of marginal probabilityare basically ingood agreement, which proves the correctness of the method in this paper.

5. Conclusions and FurtherStudies

In this paper, the FPK equation and steady FPK equation are obtained from the stochastic model of a SMIB system with stochastic excitationof renewable energy, and the steady probability density of the system under random excitation of different intensities is analysed by numerical simulation. It can be captured from the results that the value of power angle is concentrated in the concentration trend weakens when the random excitation intensity of the system increases. The follow-up work is to study the evolution of probability density with time variable $t$and the stochastic bifurcation problem of the system as the parameter changes.

The result herein gives us a clueto study dynamic behaviour of power systems with renewables. The method proposed here can be used to analyse multi-machine system. Apply the method given in [19-20], we can first divide the generators into two groups, and then get the equivalent two-machinesystem. After further simplification, the multiple-machine system will be equivalent to a SMIB system, whichcan be used to obtain the FPK equation and calculate the numerical solution according to the preceding method.

References

[1] Jianyong Z, Ping J, Yiping Y and Feng W 2012 *Sci. China Technol.* **55** pp 1873-80
[2] Xiaoshu L and Duqu W 2015 *Nonlinear dynamical behaviour and control of complex motors and power systems* (Beijing: Beijing Science Press) chapter 6 pp 122-70
[3] Yangzan H, Zengyin W2011 *Power system analysis* (Wuhan: Huazhong University of Science and Technology Press) pp 52-62
[4] Federico M and Rafael Z M 2013 *IEEE Transactions on Power Systems* **28** pp 4537-44
[5] Ming Z, Bo Y, Xiaoping Z and Gengin L 2014 *Proc. of the CSEE* **34** pp 1575-82
[6] Keyou W and Mariesa L C 2013 *IEEE Transactions on Power Systems* **28** pp 2994-3001
[7] Desmond J H 2001 *Society for Industrial and Applied Mathematics* **43** pp 525-46
[8] Zhaoyang D, Junhua Z and Divid J H 2012 *IEEE Transactions on Power Systems* **27** pp 1741-49
[9] Weiqiu Z 2006 *Applied Mechanics Reviews* **59** pp 230-48
[10] Lawrence C E 2014 *An Introduction to Stochastic Differential Equations* (New York: American Mathematical Society)
[11] Xuerong M 2007 *Stochastic Differential Equations and Applications* (Elsevier Science & Technology, Horwood Publishing Ltd) Chapter 2 pp 47-56
[12] Qiming W 2006 Research on Ito stochastic differential equation and its application in electric power systems *Master Thesis Hohai University, Nanjing*
[13] Zhilong H 2006 Research on dynamics and control of several nonlinear stochastic systems *Ph.D. Thesis, Zhejiang University, Zhejiang*
[14] Zhilong H and LiqiangZ 2003 *Chinese Conf. Computational Mechanics (CCCM)*pp 165-71
[15] Pichler L, Masud A and Bergman L 2013 Numerical solution of the Fokker–Planck Equation by finite difference and finite element methods—A comparative study*(Computational Methods on Stochatic Dynamicsvol2)* edManolis P, George S and Vissarion P (New York: Springer) chapter 5 pp 69-76
[16] Ping J, Hongyu L, Yusheng X, Yongfei L and Feng W 2013 *J. Hohai University(Natural Sciences)* **6** pp 536-41
[17] Claudio F 2013 *Engineering* **5** pp 975-88

[18] Qun L 2017 *Basic theory of numerical solution of differential equations* (Beijing: Tsinghua University Press) chapter 2 pp 112-31
[19] Hongyu L, Ping J, Yiping Y, Xiaoming H and Xinqi C 2015 *Proc. of the CSEE* **35** pp 3561-68
[20] Hongyu L, Ping J, Xinqi C 2015 *Sci. China Technol.* **45** pp 766–72

Acknowledgements
This work was supported in part by National Natural Science Foundation of China (NSFC, Grant No.51207052) and the Fundamental Research Funds for the Central Universities (Grant No. 2014MS62).

IOP Conf. Series: Journal of Physics: Conf. Series **1016** (2018) 012022 doi:10.1088/1742-6596/1016/1/012022

Research of PV Power Generation MPPT based on GA-BP Neural Network

Yu Su[1], Xianfu Lin[2]

1Beijing Institute of Technology,Zhuhai, School of Information Technology, Guangdong Zhuihai,519088,P.R.China
2Central South University, School of Information Science and Engineering, Hunan Changsha, 410083,P.R.China

E-mail: 2007youter@163.com

Abstract. Photovoltaic power generation has become the main research direction of new energy power generation. But high investment and low efficiency of photovoltaic industry arouse concern in some extent. So maximum power point tracking of photovoltaic power generation has been a popular study point.Due to slow response, oscillation at maximum power point and low precision, the algorithm based on genetic algorithm combined with BP neural network are designed detailedly in this paper.And the modeling and simulation are completed by use of MATLAB/SIMULINK. The results show that the algorithm is effective and the maximum power point can be tracked accurately and quickly.

1.Introduction

Nowadays energy shortage and environmental pollution problems are attracting attention. Solar energy is one kind of widely used renewable energy in the current availability renewable energies because it is abundant, clean and safe. So many countries have increased the investment of photovoltaic research. The photovoltaic array is easily influenced by the light and temperature of the environment, and output characteristics have obvious nonlinear. Therefore solar energy photovoltaic power cost is high and its conversion efficiency is low. In order to solve the problem, it is important to carry out the maximum power point tracking.[1,2] A variety of commonly used maximum power point tracking methods are introduced at home and abroad. [3,4,5]Analyzing the conventional MPPT control methods, the advantages and disadvantages are listed.BP neural network also has strong nonlinear approximation ability, but its gradient decent algorithm determines it is easy falling into local optimum and sensitive to the initial values. So the algorithm based on genetic algorithm combined with BP neural network are proposed in this paper for carrying out the maximum power point tracking.Improving the efficiency of photovoltaic array is a great sense to the development of photovoltaic industry.

2.Analysis of the output characteristics of photovoltaic array

The output characteristics of photovoltaic array refer to relationships under certain temperature and light intensity. One is the relationship between the output voltage and current of the battery, the other is the output voltage and the corresponding power relation.According to the theory of Electronics[6], the equivalent circuit of photovoltaic array is shown in figure 1:

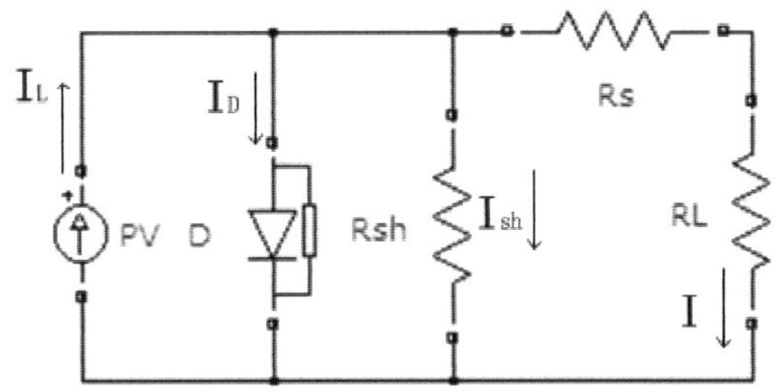

Figure 1. equivalent circuit of photovoltaic array

The parameters of one existing photovoltaic array are as follows:

$$U_{oc} = 350V, I_{sc} = 13A, U_m = 260V, I_m = 10A$$

In the condition of illumination intensity Gref=1000W/m2 ,the $I-U$ characteristic curve and $P-U$ characteristic curve are obtained by using MATLAB/SIMULINK and shown in Figure 2 ,when temperature is set to $0^o C$、$25^o C$、$50^o C$、$75^o C$、$100^o C$. At the same time in the condition of temperature $T = 25^o C$, the $I-U$ characteristic curve and the $P-U$ characteristic curve are shown in Figure 3, when the illumination intensity is set to $200W/m^2$、$400W/m^2$、$600W/m^2$、$800W/m^2$、$1000W/m^2$.

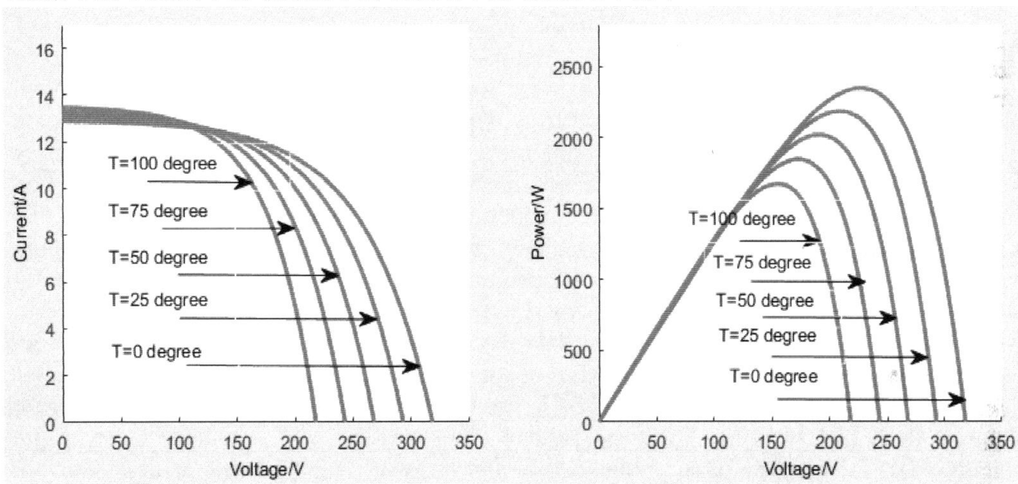

Figure 2. I-U (left) characteristic curve and P-U characteristic curve (right)

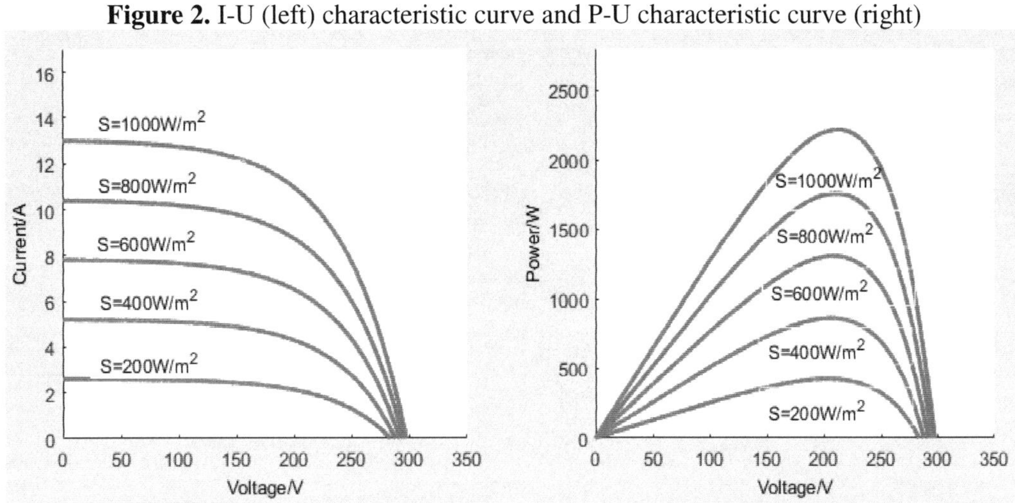

Figure 3. I-U (left) characteristic curve and P-U characteristic curve (right)

By analyzing the above curves, it is obvious that the output characteristics of photovoltaic array are nonlinear. And with the change of environmental factors the maximum power point will inevitably change. In order to improve photovoltaic conversion efficiency, it is very necessary to make the photovoltaic array work near the maximum power point.

3.MPPT analysis of GA-BP neural network

3.1.Analysis of advantages and disadvantages of BP neural network

BP neural network is a multilayer feedforward neural network with error back propagation algorithm[7], which is one of the most widely used neural network models in the present application. BP neural network with the ability of nonlinear mapping implements mapping function from input to output. Its three layer neural network can approximate any nonlinear continuous function with any precision and has a high degree of self-learning and self-adaptation. However, its disadvantages are also quite obvious. (1)The local minimization problem that is different initial weights will lead to the network into local extremum point or non convergence. (2)Slow convergence. (3)Because the initial weights and thresholds of the training network are randomly generated, which are lack of basis for choice.But the final imitative effect of data is determined by weights and thresholds.The choice of initial weight determines the direction of convergence. So Its prediction ability is not ideal.

3.2.Genetic algorithm and BP neural network

The genetic algorithm combined with BP neural network for making up the deficiency of the existing problems is designed. In this paper, we use this algorithm to establish the maximum power point tracking of photovoltaic array. The processes are as follows:(1)Initialization parameters of BP neural network.Because the maximum power point of the photovoltaic array is affected by light and temperature, so the number of neurons in the input layer is 2, and by trial and error the number of neurons in the hidden layer is set to 5, the output neuron is 1 that is the maximum power point.(2)Initial population and population coding.Taking into account the tracking accuracy of the maximum power point and reducing non-essential process of encoding and decoding, it adopts floating point real number as gene coding.(3)The error between the forecast data and the expected data is selected as fitness function.(4)Loop selection, crossover, mutation, and computational adaptation need to be done until evolution times is reached.In this selection roulette method is used for individual choices.(5)Getting the best initial weights and thresholds.(6)Using genetic algorithm to work out the optimal initial weights and thresholds for training BP neural network.The GA-BP neural network architecture is shown in Figure 4 and diagram flow chart is shown in Figure 5.

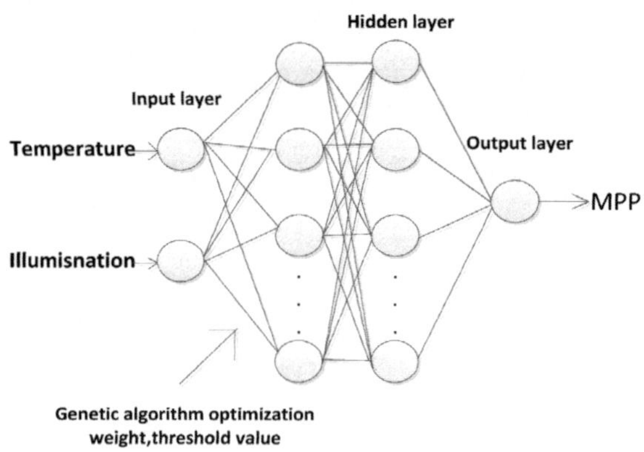

Figure 4. GA-BP neural network architecture

IOP Conf. Series: Journal of Physics: Conf. Series **1016** (2018) 012022 doi:10.1088/1742-6596/1016/1/012022

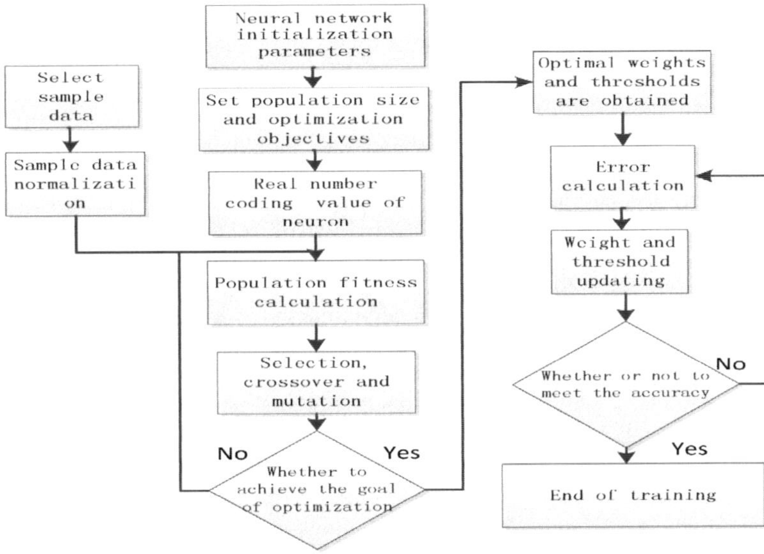

Figure 5. Algorithm flow chart

3.3.Training of GA-BP neural network

Firstly, the 500 sets of data are obtained through the P-U output characteristics of PV array,which is the maximum power point data at different temperature and illumination intensity. Among them 350 sets of data are training samples and 150 sets of data are test samples.Each set of data includes three values (temperature, illumination intensity, maximum power point). As to initial parameters of BP training network, Input layer transfer function adopts tansig type,output layer function adopts purelin type,the training function adopts trainlm, and expected error is set to 0.00015, learning rate is set to 0.02, maximum of training steps is set to 2000.For the weight and threshold of genetic algorithm initial parameter values are selected as follows,population size is 200,genetic algebra is 150,crossover rate is 0.3, and mutation rate is 0.2.By use of MATLAB/SIMULINK the PV array MPPT model is shown in figure 6.

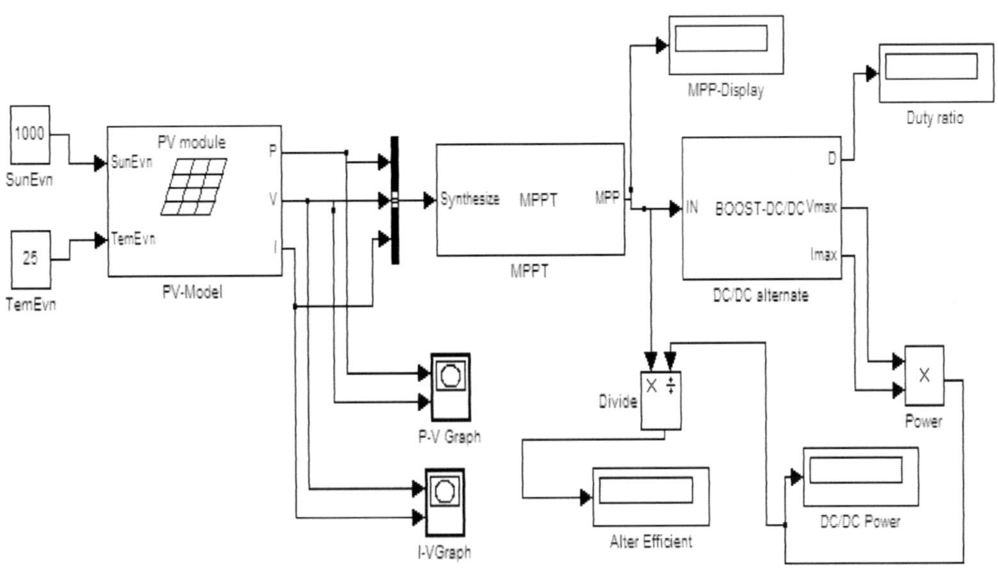

Figure 6. Simulation model of photovoltaic array MPPT

IOP Conf. Series: Journal of Physics: Conf. Series **1016** (2018) 012022 doi:10.1088/1742-6596/1016/1/012022

The main work is as follows. During runtime, the temperature and illumination intensity is modified. Then the GA-BP neural network which has been trained starts to track the maximum power point.Through the BOOST voltage module, the duty cycle is adjusted. Thus changing the output voltage of the photovoltaic array and working on the maximum power point tracked by the GA-BP neural network is implemented. The iteration process respectively used GA-BP neural network (left) and BP neural network (right) is shown in figure 7.From this figure, under the same conditions we can find that the number of iterations is only 8 by use of genetic algorithm combined with BP neural network. Compared with 58 times only using BP neural network algorithm, the convergence rate is greatly improved. The two different kinds of algorithms for tracking maximum power of photovoltaic array is shown in figure 8. From this figure, we can find that the tracking by use of BP neural network has a better accuracy in the condition of $T = 25^{o}C$, G from $800W/m^2$ to $G = 1000W/m^2$.But in other cases the algorithm will bring on higher error. By use of GA-BP neural network, in most cases it can achieve high accuracy of the PV array maximum power point tracking. In addition it can be seen from Figure 9, if using GA-BP neural network algorithm to track maximum power point, the error is about 2%. Compared with only using BP neural network algorithm, the error is reduced about 3%.

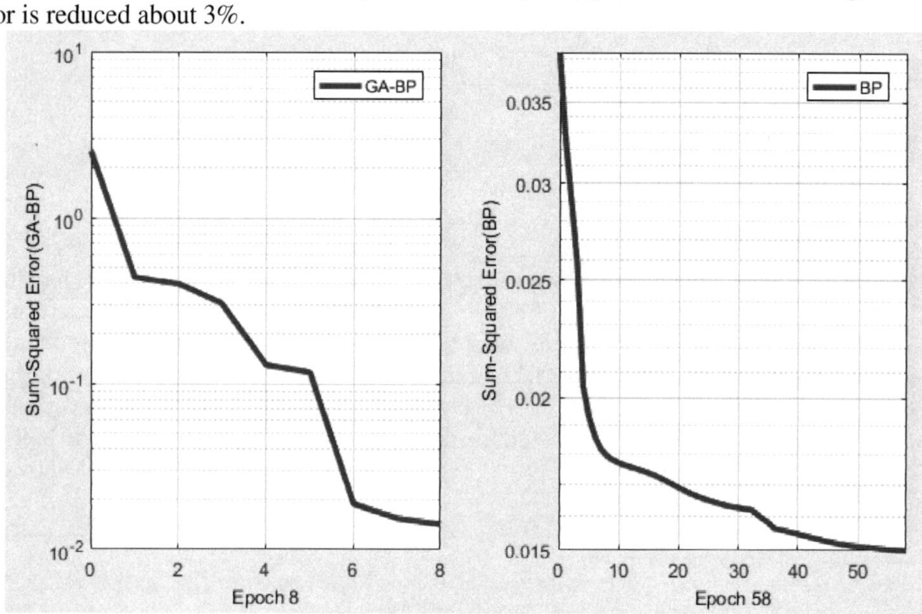

Figure 7. Iteration process

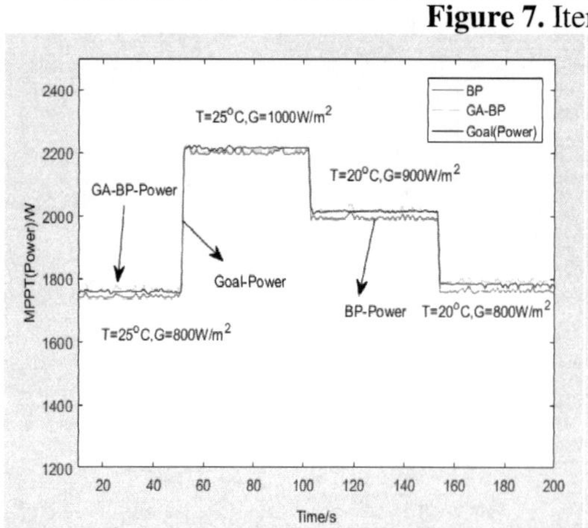

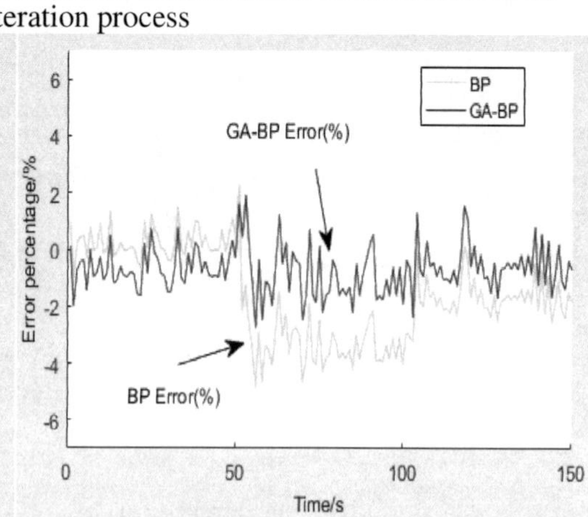

Figure 8. MPPT comparison of BP and GA-BP **Figure 9:** MPPT error of BP and GA-BP

4. Summary

The convergence rate of the BP neural network for tracking the maximum power has its own characteristics, such as slow, low accuracy and local optimal value. The efficiency of photovoltaic array can not be ignored. In order to optimize, the proposal of genetic algorithm combined with BP neural network is designed. In this paper, we analyze the $P-U$, $I-U$ output characteristics of photovoltaic arrays firstly. Due to the maximum power point of PV arrays is dynamic and changes with different external conditions. The genetic algorithm combined with BP neural network is used for optimizing and tracking the maximum power point in different cases. The algorithm is introduced in this paper, at the same time its modeling and simulation by MATLAB/SIMULINK are completed in detail. The simulation results show that the algorithm not only improves the convergence rate, but also reduces the relative error. The optimization is effective and feasible, and the maximum power point can be tracked accurately and quickly.

References

[1] Chen K Feng L Jia L Z and Li G J 2013 Research on local maximum power point tracking algorithm for photovoltaic system under local shading condition *Renewable Energy Resources* 13-16
[2] Zhang H Y Yan YU and Xu K 2015 Maximum power tracking strategy of distributed PV generation based on PSO *Control Engineering* 51-57
[3] Zhang Y X Li X and Zhao J 2013 Improved disturbance observation method for maximum power tracking of photovoltaic power source *Journal of Tianjin University of Technology* 21-23
[4] Wang Y L Yang X H Wang J C Wang Y L Li H R and Feng C C 2015 A new variable step size photovoltaic maximum power point tracking control strategy *Electric Machines and Control* 54-57
[5] Gong M F Li J and Li Z 2016 Research of photovoltaic array based on Simulink MATLAB *Test and Measurement Technology* 15-20
[6] Wu C J 2015 Design of power control for photovoltaic power supply based on Neural Network *Research and Design Power Technology* 147-149
[7] Ren Z W and San Y 2007 Improvement and Performance Research of real number genetic algorithm *Electronic journal* 269-275

Acknowledgement
This research was financially supported by the Guangdong Province Education Office Foundation for excellent young teachers 2015.